Keratocyte Induced Corneal Microedema (KME)

A Novel Pathophysiologic Model describing the Etiology, Physiology and Management of Diffuse Lamellar Keratitis, Central Toxic Keratopathy, Central Flap Necrosis, Flap Necrosis Syndrome and Central Lamellar Keratitis

Brian R. Will, MD
Adjunct Clinical Professor of Ophthalmology
Department of Ophthalmology
Loma Linda University School of Medicine
Loma Linda, California

Hisham Adel Hasby Saad, MD
Assistant Lecturer of Ophthalmology
Faculty of Medicine, Tanta University
Tanta, Egypt

Samuel Lee MD and David Skale MD
Department of Ophthalmology
Loma Linda University School of Medicine
Loma Linda, California

Foreward

"Unthinking respect for authority is the greatest enemy of truth"
Albert Einstein

Unfortunately, a critical analysis of current concepts promoted in the peer and non-peer literature on the etiology of Diffuse Lamellar Keratitis (DLK), Central Toxic Keratopathy (CTK), Central Flap Necrosis (CFN), Flap Necrosis Syndrome (FNS) and Central Lamellar Keratitis (CLK) reveals that many of these broadly held beliefs are not consistent with known physiologic principles, frequently appear to promote belief systems that lack scientific validation and on occasion seem to even invoke magical thinking. There is also a troubling propensity towards a reliance on anecdotal data and a frequent capitulation to flawed logic. A fundamentally flawed interpretation of clinical and basic science evidence is the norm rather than the exception.

My reluctant journey into the chaotic abyss of the confused, muddled and convoluted "science" characteristic of public discourse on these disorders abruptly began on a beautiful August day in 2002 when a young school teacher presented with bilateral Stage 4 DLK and vision of 20/400. From the crucible of that ordeal, further amplified by the traumatic experience of the five other patients that followed in her wake, I came to recognize that the pathophysiology and etiology for Stage 4 DLK as defined by every textbook, journal and scientific session on LASIK was completely incorrect.

If the map doesn't agree with the ground, the map is wrong
G Livingston MD

In my search to understand these patients, I began to realize that a more systematic and reasoned approach was needed if we were to understand this paradox. A pivotal moment came when I received a random email that pointed me to a basic science paper on interstitial fluid pressures (Pif) coauthored by Stephen Klyce in IOVS. As a medical student at the Loma Linda University School of Medicine, I had spent the summer of 1983 in the Perinatal Biology section performing research seeking to understand the control mechanism for Pif in the body. Although we ultimately published our results in the American Journal of Physiology, I had long forgotten the implications of that work. However, the paper by Dr. Klyce's group struck a resonant chord that was stunning in its application. I quickly began to realize that control of interstitial fluid pressures in the cornea was at the core of understanding the pathogenesis of DLK, Stage 4 DLK, CTK, CFN, FNS and CLK. Unfortunately, the discovery of multiple catastrophic errors in current physiologic paradigms was only the beginning. Redefining a pathophysiologic map that rationally and comprehensively described the etiology and mechanism of these disorders has been an arduous and largely ongoing task.

"All truth passes through three stages:
First, it is ridiculed
Second, it is violently opposed; and
Third, it is accepted as self evident"
Arthur Schopenhauer (1788-1860)

The objective of directly challenging conventional scientific dogma on the topic of DLK and related entities is foolhardy and insane at best. Unfortunately, after rigorous review and study, it is overwhelmingly clear that the clinical and basic science data refuting conventionally accepted clinical paradigms is unequivocal and our OCT and Pentacam images are irrefutable. Moreover, more and more bright Ophthalmic surgeons have begun to question the validity of current disease paradigms. Unfortunately, the complexity of these topics combined with the profound ignorance amongst Ophthalmologists regarding the most basic of physiologic principles makes communication of these ideas in the context of a five minute paper presentation or a ten page journal paper nearly impossible.

If the task was not absurd enough already, the pathophysiologic model that we propose not only challenges current dogma in refractive surgery on the topic of DLK, but attempts to simultaneously refute four (4) broadly held paradigms in Ophthalmology including:

a. The etiology and pathophysiology of DLK
b. The etiology and pathophysiology of CTK
c. The mechanism that defines control of interstitial fluid pressures in the cornea
d. The primary etiology and control of corneal biomechanics

The profoundly revolutionary concepts articulated in this textbook redefine nearly everything we 'know' about control of fluid dynamics and the biomechanical forces that shape in the cornea. This is particularly true with respect to understanding how and why the cornea responds to laser refractive surgery, collagen crosslinking and femtosecond laser photodisruption and the etiology of disorders such as keratoconus.

Despite my personal impression that it would be simpler to explain "String Theory" to people from the Dark Ages who cling to a belief in a flat, earth centric universe than to explain Keratocyte Induced Corneal MicroEdema to refractive surgeons, my journey will haltingly begin with this pitiful textbook.

If the ideas in this book can be shown to be in error, I would gratefully abandon them and rejoin my many colleagues that appear to believe in myth and magic. However, absent such a rebuttal, with my professional reputation at risk, I follow in the footsteps of Martin Luther who declared before the Diet of Worms - "Here I stand; I can do no other. God help me. Amen."

Dedication

To all those who, at their peril, ask Why?

To my devoted wife Carlene without whose unconditional love, support and often spirited reproof, this book would never exist

To my youngest son Nicholas who, tired of my whining over the "abject stupidity" of my peers, cajoled me into putting pen to paper to create my "Nobel prize winning" thesis

To the doctors who consistently present papers, posters and write textbooks filled with gross errors on these topics whose profound ignorance and arrogance finally drove me to the brink of lunacy. I can take no more. Possessed with an obsession for defensible science and encumbered by the disease of verbosity, I set my sail into the wind.

Keratocyte Induced Corneal MicroEdema (KME)

Table of Contents

Keratocyte Induced Corneal MicroEdema (KME)

The Current Debacle

The pathogenesis of Diffuse Lamellar Keratitis (DLK) is generally considered to be an antibody mediated inflammatory reaction to an antigen introduced into the LASIK interface at the time of surgery[1-6].

> Perhaps no other topic in the field of refractive surgery is as bereft of scientific discipline as the clinical constellation of Diffuse Lamellar Keratitis, Stage 4 Diffuse Lamellar Keratitis, Central Toxic Keratopathy, Central Flap Necrosis, Flap Necrosis Syndrome, Central Lamellar Keratitis and Interface Fluid Syndrome

This inflammation is believed to lead to a build up of inflammatory cells in the interface that can result in the release of collagenases and matrix metalloproteinases that cause central stromal necrosis[4,7-9]. A broad spectrum of substances have been described as potential antigens[2,4,7-11]. Although this mechanism may account for epidemic reports of DLK, it does not fully account for all cases[6,8,12,13]. Recent studies suggest that molecular messengers, proinflammatory mediators and chemokines that regulate the corneal wound healing response may be implicated in the pathophysiologic cascade for DLK[7,14-21]. Although self-limited if the degree of inflammation is mild or moderate, severe inflammation (Stage 4 DLK) is believed to cause necrosis, volume loss, flap macrostriae and potential reduction in best corrected visual acuity (BCVA)[1,4-6].

Central Toxic Keratopathy (CTK) is described as a toxic response of the central stroma to an unknown toxin[22]. This toxin is believed to cause central stromal necrosis, volume loss, flap macrostriae and potential reduction in BCVA although, with healing, the patient may not have permanent visual deficits. The toxin, its action on corneal tissue and the mechanism of healing are currently unknown but this disorder is specifically noted to occur in the apparent absence of a significant acute inflammatory response.

Controversy exists as to whether Stage 4 DLK represents the terminal state of a progression, or whether it is a separate entity from the milder stages[22]. Additional debate revolves around whether Stage 4 DLK and CTK are variations of the same disorder[23].

> The series of cases presented in this paper demonstrate characteristics that are distinctly different from all of these entities and further add to this controversy

In addition, various authors have described entities currently identified as Central Flap Necrosis (CFN)[24], Flap Necrosis Syndrome (FNS)[25] and Central Lamellar Keratitis (CLK)[26] that appear to exhibit similar characteristics to Stage 4 DLK and CTK. The series of cases

The cases in this consecutive series of eyes affected by Stage 4 DLK illustrate several key features that are in conflict with current concepts with respect to the pathophysiology of this disorder.

1. All patients recovered a BSCVA of 20/20 or better
2. At the completion of therapy, no patient had residual glare, halos, distorted vision or other adverse clinical complaint
3. All eyes demonstrated marked hyperopic shifts combined with significant levels of induced astigmatism that were rapidly reversed using topical hyperosmotic agents
4. Only 50% of the eyes required laser enhancement and no eyes required customized wavefront or topographically driven enhancement to attain an acceptable visual outcome
5. In all cases, topical steroid dosing was discontinued at presentation of Stage 4 DLK thereby markedly diminishing consideration of their value in promoting disease recovery
6. Recovery to BSCVA of 20/20 or better was accomplished within a few weeks or months of the initial onset of Stage 4 DLK
7. All eyes in this series exhibited severe pathology isolated to the central and paracentral anterior cornea that markedly adversely affected BSCVA

8. At the time of flap lift and irrigation procedures the tissue of the stromal bed and flap was normal in all cases that underwent that procedure
9. The degree of opacification of the central cornea appears to be markedly different when comparing slit lamp examination with direct observation techniques with a penlight

presented in this paper demonstrate characteristics that are distinctly different from all of these entities and further add to this controversy.

Rather than using this case series to define

In order to understand the relationship between the cases in this series and DLK, CTK, CFN, FNS or CLK, it is essential to develop a comprehensive paradigm that describes the etiology and pathophysiology of these disorders. We posit that all current paradigms are fundamentally flawed

a new disease or disease variant, we believe that in order to understand the relationship between the cases in this series and DLK, CTK, CFN, FNS or CLK, it is essential to develop a comprehensive paradigm that describes the etiology and pathophysiology of these disorders. Armed with such a paradigm we can develop a more reasoned classification system and define best management practices for DLK, CTK, CFN, FNS and CLK.

We present a consecutive series of eyes demonstrating Stage 4 DLK that recovered full visual function following completion of therapy. This unique series

> If the map doesn't agree with the ground, the map is wrong
> G Livingston MD

has been previously partially presented at the Aspen Invitational Refractive Surgery Symposium in 2003 (Aspen, Colorado, USA) with the entire series being presented at the American Society of Cataract and Refractive Surgery in 2005 (Chicago, Illinois, USA).

We further propose a novel mechanism that describes a common pathophysiologic process for DLK, Stage 4 DLK, CTK, CFN, FNS and CLK. We further discuss the implications of this model on strategies for prevention and management of this disease process.

> We reject all current paradigms and propose a novel mechanism that describes a common pathophysiologic process for DLK, Stage 4 DLK, CTK, CFN, FNS and CLK

The Clinical Paradox

Of primary concern is that these clinical findings are nearly impossible to explain in the context of disease paradigms that invoke tissue necrosis as a core aspect of the pathophysiology. If tissue necrosis caused by lytic enzymes released by inflammatory cells is the primary underlying cause of this disorder, logically the likelihood that 100% of eyes in a consecutive series of all eyes with Stage 4 DLK would recover full visual acuity without adverse clinical effects is extraordinarily improbable

We present a consecutive series of 8 eyes of 6 patients over a 21 month interval between August, 2002 and April, 2004 that developed central opacities and interface infiltrates consistent with Stage 4 DLK following LASIK. Cases were analyzed by retrospective review of medical records. All patients had manifest refraction, keratometry, corneal topography, pachymetry, slit-lamp examination, and fundoscopic examination prior to surgery. The eyelids in all cases were prepared with 70% isopropyl alcohol swab sticks. No Betadine® (Purdue Pharma LP, Stamford, CT, USA) was used at any stage in any of these cases. Preoperatively patients received pH neutral, chilled Proparacaine 0.5% (Bausch and Lomb, Rochester, NY, USA), preservative free Naphcon-A (Leiter's Pharmacy, San Jose, CA, USA) and Acular® PF (Allergan Inc., Irvine, CA, USA).

All flaps were created in an uncomplicated manner using the Intralase™ FS laser (Advanced Medical Optics, Santa Ana, CA, USA) operating at 15 kHz with an intended flap thickness of 90-100 microns and a nasal hinge, raster energy ranging from 2.5 to 2.6 microjoules, a 10 micron spot and line separation and a side cut energy ranging between 4.0 and 5.2 microjoules. All cases underwent uncomplicated non-wavefront excimer laser treatment of varying depths and optical zones. The first patient underwent treatment with a LADARVision® excimer laser (Alcon, Fort Worth, TX, USA) and the remaining 5 with the VISX™ Star S4 (Advanced Medical Optics, Santa Ana, CA, USA).

As a group, the eyes in this series are relatively diverse. In addition, the clinical approach to case management evolved over time as previous cases provided valuable experience with respect to what aspect of our therapy appeared to be

The clinical outcome for all eyes was excellent

beneficial for accelerating recovery and what did not. Despite these limitations, the clinical outcome for all eyes was excellent.

Overall, the series represents 8 eyes of 6 patients of which 4 were male and 2 were female. Three of the eyes were right eyes and 5 were left eyes. The age of the patients ranged from 32 to 43 with a mean of 37.1 years. Six eyes were myopic. Of those, preoperatively 4 eyes demonstrated high myopia, 1 eye demonstrated moderate myopia and 1 eye exhibited a high degree of regular astigmatism. The remaining two eyes were mixed astigmats.

> All eyes recovered best spectacle corrected visual acuity (BSCVA) of 20/20 or better. No patient had complaints of untoward visual disturbances at completion of therapy. No patients exhibited residual corneal scarring or visually significant flap microstriae

Presentation of Stage 4 DLK in this series ranged from the 5th to the 10th day postoperatively from the primary LASIK procedure. BCVA at time of presentation ranged from 20/50 to 20/400. All eyes exhibited a marked hyperopic shift whose spherical equivalent (SE) ranged from +1.00 to +3.75 D and a mean of +2.64 D. Seven of the 8 eyes exhibited induced astigmatism ranging from +0.25 to +1.75 D with a mean of +1.1 D.

Only 4 of 8 eyes underwent immediate flap lift and rinse either prior to or at time of their presentation of Stage 4 DLK, as is generally prescribed in the management of DLK. The remaining 4 eyes did not undergo an immediate flap lift and irrigation procedure. To remove microstriae, several weeks following the primary LASIK procedure 4 of 8 eyes underwent manual removal of the central epithelium accompanied by flap stretching while one eye underwent flap lift and stretch without epithelial removal. In all, 7 of 8 eyes underwent some form of flap lift procedure. No eyes received sutures for flap replacement. In each of the 7 eyes that underwent flap lift, the stromal bed and flap tissue were judged to be normal and there was no evidence of tissue necrosis.

Only 4 of 8 eyes underwent laser enhancement for residual refractive error using standard flap lift techniques. Two of those eyes underwent 2 laser enhancements for a large hyperopic shift. All laser enhancements were performed on a VISX™ Star S4 excimer laser using conventional treatment profiles. No patients received wavefront or topography guided photoablations. Four of 8 eyes received no additional laser treatment.

All eyes recovered best spectacle corrected visual acuity (BSCVA) of 20/20 or better. No patient had complaints of untoward visual disturbances at completion of therapy. No patients exhibited residual corneal scarring or visually significant flap microstriae. All patients were discharged from the clinic at the conclusion of their treatment program. No patients were lost to follow-up.

> These clinical findings are nearly impossible to explain in the context of disease paradigms that invoke tissue necrosis as a core aspect of the pathophysiology

Of primary concern is that these clinical findings, for reasons we will define in this paper, are nearly impossible to explain in the context of disease paradigms that invoke tissue necrosis as a core aspect of the pathophysiology. If tissue necrosis caused by lytic enzymes released by inflammatory cells is the primary underlying cause of this disorder, logically the likelihood that 100% of eyes in a consecutive series of all eyes with Stage 4 DLK would recover full visual acuity without adverse clinical effects is extremely small and extraordinarily improbable.

In the context of this unique patient series, combined with a review of all pertinent scientific literature, we propose a novel pathophysiologic mechanism that helps explain this clinical paradox.

> We propose a novel pathophysiologic mechanism that helps explain this clinical paradox

Table I.

Patient demographic information, preoperative parameters, presentation characteristics, treatment protocols and refractive outcomes are provided. All eyes demonstrated an initial marked loss of BSCVA, interface inflammation, flap striae and a hyperopic shift. With management, all eyes recovered BSCVA of 20/20 or better with no adverse visual complaints. Only four eyes underwent excimer laser enhancement using non-custom ablation protocols.

Patient	Case 1	Case 2	Case 3	Case 4
Age	34	34	32	43
Gender	F	F	M	M
Eye	OD	OS	OD	OD
Manifest Rx	-9.5	-7.75	-0.75+1.00X7	-8.25+0.25X157
Preop BCVA	20/15	20/15	20/15	20/15
Presentation	Day 5	Day 5	Day 10	Day 5
BCVA	20/60	20/400	20/200	20/400
Lift + Rinse	No	Yes	No	No
Flap Stretch	No	No	Yes	Yes
Topical Steroid	No	No	No	No
Hyperosmotics	Yes	Yes	Yes	Yes
Topical NSAID	No	No	Yes	Yes
Oral Vit C	No	No	No	No
Max Hyper Shift	+3.00+1.50X30	+1.50+1.75X65	+3.00+0.75X169	+1.00
Laser Enhance	No	Yes	Yes X 2	No
Final BSCVA	20/20	20/20	20/20	20/15
Adverse Complaints	None	None	None	None

Patient	Case 5	Case 6	Case 7	Case 8
Age	43	36	35	43
Gender	M	M	M	F
Eye	OS	OS	OS	OS
Manifest Rx	-7.75+0.50X65	-0.50+1.25X97	-4.25+3.00X93	-4.25
Preop BCVA	20/20	20/20	20/15	20/15
Presentation	Day 5	Day 8	Day 6	Day 6
BCVA	20/200	20/100	20/200	20/50
Lift + Rinse	No	Yes	Yes X 2	Yes
Flap Stretch	Yes	No	Yes	Yes
Topical Steroid	No	No	No	No
Hyperosmotics	Yes	Yes	Yes	Yes
Topical NSAID	Yes	Yes	Yes	Yes
Oral Vit C	No	No	No	Yes
Max Hyper Shift	+2.00+1.75X105	+3.00+1.00X37	+2.50+0.75X47	+1.25+0.25X42
Laser Enhance	No	Yes X 2	Yes	No
Final BSCVA	20/20	20/20	20/20	20/20
Adverse Complaints	None	None	None	None

Enigmatic Nature of The Paradigm Lost

Fatally Flawed Science

The cases described in this consecutive series of eyes demonstrate a number of characteristics that are distinctly at odds with previously published observations of

> We expose several fundamental flaws in widely accepted concepts regarding the pathogenesis of DLK, Stage 4 DLK, CTK, CFN, FNS and CLK including:
>
> - Lack of consistency with accepted physiologic principles
> - Promotion of belief systems that lack scientific validation
> - Use of magical thinking
> - Reliance on anecdotal data
> - Application of flawed logic

Stage 4 DLK. Of additional concern is that these differences not readily explained using commonly accepted pathophysiologic explanations for the etiology or mechanism of Stage 4 DLK, CTK, CFN, FNS or CLK. This is highly problematic. However, despite these marked incongruities, we do not believe that our cases represent some new disease or variant of these previously described entities.

Further, it is our perspective that the clinical findings of these patients appear to expose several fundamental flaws in widely accepted concepts regarding the pathogenesis of DLK, Stage 4 DLK, CTK,

CFN, FNS and CLK. Our findings also call into question the viability of our current classification system for DLK and its related entities of CTK, CFN, FNS and CLK. In that context, we believe it is reasonable to pose the question as to whether such entities actually represent distinctive diseases versus variants of a single common corneal disorder.

> As a consequence of these challenges, there is a lack of a unifying pathophysiologic theme for these disorders. In the absence of such a comprehensive model, the apparent epidemic and sporadic characteristics of these entities has confounded our ability to identify specific etiologies and has confused best practices in disease management and prevention

In order to provide a systematic and reasoned approach to understanding these disorders, it is our intent to develop a comprehensive model to better define the pathogenesis of DLK, Stage 4 DLK, CTK, CFN, FNS and CLK. Unfortunately, a critical analysis of the current concepts promoted in the peer and non-peer literature on the etiology of these entities suggests that many times such ideas are not consistent with known physiologic principles, frequently appear to promote belief systems that lack scientific validation and on occasion seem to even invoke magical thinking. This review also reveals a troubling propensity towards a reliance on anecdotal data and a frequent capitulation to flawed logic. As a consequence of these challenges, there is a lack of a unifying pathophysiologic theme for these disorders. In the absence of such a comprehensive model, the apparent epidemic and sporadic characteristics of these entities has confounded our ability to identify specific etiologies and has confused best practices in disease management and prevention.

Fundamental Challenges to Current Pathophysiologic Paradigms

With the goal of developing a pathophysiologic model, it is our initial purpose to methodically evaluate these inconsistencies in order to provide a systematic and comprehensive review of these entities.

Challenges to current paradigms include an inability to explain:

1. Why the major pathology is always located centrally
2. Why inflammatory cells in the interface move centrally
3. Why a hyperopic shift is commonly observed in these disorders
4. Why the hyperopic shift is frequently reversible
5. Why intraoperative observations are not consistent with tissue necrosis
6. Why no apparent tissue necrosis occurs in Stage 1 – 3 DLK
7. Why there is no apparent continuum of effect
8. Why the inflammatory cells assume a "Shifting Sands" appearance
9. Why Stage 4 DLK and CTK in many cases can be "self limited" leaving no residual tissue scar or adverse refractive effect
10. Why there have been no published reports of corneal transplantation for Stage 4 DLK in over 10 years
11. Why epidemic, sporadic and late onset cases have no common etiologic pathway
12. Why current classification systems fail as a clinically relevant tool
13. Why published histopathology studies are not consistent with current theories on etiology
14. Why published pachymetry data fails to support current theory
15. What exact role inflammatory cells play in the pathogenesis of these disorders

Why is the major pathology exclusively located centrally in the cornea?

Perhaps the most striking observation with respect to Stage 4 DLK, CTK, CFN, FNS and CLK is the fact that in 100% of published cases, including the 8 cases in our consecutive series, the major pathology commonly attributed to tissue necrosis is exclusively located in the center of the cornea[2,4-6,24-26]. In view of current theory, the exclusive localization of the primary pathology in the central cornea is a serious challenge to explain for any of these disorders.

> The exclusive localization of the primary pathology in the central cornea is a serious challenge to explain for any of these disorders. Given the spectrum of antigens currently believed to cause DLK and absence of a viable 'toxin' for CTK, this type of non-random antigen distribution is highly unlikely

If one adopts the position that acute inflammatory cells in the interface are the cause of the central necrosis in Stage 4 DLK[4-8] then these cells must either be drawn to the corneal center[2,4] or navigate there through other mechanisms. If these cells are drawn to the corneal center via the mechanism of chemotaxis then it logically follows that the specific antigen involved is predominantly or preferentially deposited in the center of the cornea in 100% of these published cases. Given the spectrum of antigens currently believed to cause DLK, this type of non-random antigen distribution is highly unlikely. However, if the deposition of the antigen is diffuse or random and the mechanism driving cell movement is chemotaxis, then the consistent directional migration of acute inflammatory cells with accumulation and "clumping" in the corneal center is similarly difficult to explain.

To further explore what we perceive to be the magical thought patterns required to accept currently promoted concepts of disease pathogenesis it is likely useful to examine the manner in which multiple antigenic compounds can be conceived to be distributed in the LASIK interface. Currently, it is proposed that lipopolysacchride endotoxin residues from sterilizers[6,10,13,15,27,28], talc or silicone contaminants from surgical gloves[29,30], meibomian secretions[5,6,31], ink from marking pens applied to the cornea or conjunctiva[32], coatings or debris on microkeratome blades[33-35], cleaning solutions for instruments[36,37], toxic chemicals[38], fluoroquinolones[39], specific types of microkeratomes[40,41], inflammatory cytokines such as platelet activating factor (PAF)[20,42], Femtosecond lasers[43], Celluvisc®[44], atopy[45-47], blood from large flaps with peripheral neovascularization[5,6], contaminated reusable irrigation cannulas[5], enhancements[6,12], blepharitis[5,6], iritis[48], aerosolized bacterial cell membranes (Mark Johnson MD, personal communication), gonococcal keratoconjunctivitis[49], merocel sponges[50] and particulate matter such as sheet rock dust in the air[51] are causes of DLK.

A comprehensive review of the literature fails to provide any guidance or predictions as to how such a litany of disparate materials could be expected to become distributed in the interface. We can conceive of no mechanism that would systematically localize such a list of contaminants, molecules or materials in the interface. The absence of any localizing process for antigenic or toxic

There is no mechanism that would systematically localize the current list of contaminants, toxins, molecules or materials proposed to cause DLK or CTK in the interface.

contaminants, combined with the nomenclature wherein this disease is referred to as a 'diffuse' lamellar keratitis, strongly suggests that the appropriate null hypothesis with respect to the distribution

of inciting factors should be that they are 'diffusely' distributed in the interface.

If we test this first null hypothesis using conventional chi-square statistical analysis using only the series of Stage 4 DLK patients in our series and ignore the additional similar cases described in the literature (even though ignoring these other cases actually makes it easier to accept a false null hypothesis) we are provided with the rather problematic conclusion summarized in Table II.

Table II - Postulating as our null hypothesis that the distribution of contaminants in the interface in DLK is diffuse, produces a statistical analysis that demonstrates that the 8 eyes in this series are highly unlikely (P<0.005) to be from such a disease process

Distribution of Pathology	Localized	Diffuse	P value
Number Observed	8	0	
Number Expected if none are localized	0	8	0
Number Expected if 20% are localized	1.6	6.4	1.54×10^{-8}
Number Expected if 50% are localized	4	4	0.0046

Postulating as our null hypothesis that the distribution of contaminants in the interface in DLK is diffuse, produces a statistical analysis that demonstrates that the 8 eyes in this series are highly unlikely (P<0.005) to be from such a disease process. If we change our definition of DLK and null hypothesis to include up to 20% or even 50% of cases where the contaminant is deposited in a highly localized manner in the interface, our null hypothesis remains false.

> In Stage 4 DLK and CTK the major pathology (apparently by definition) occur in the corneal center irrespective of the type of contaminant under consideration. This is essentially impossible to reconcile with the wide variety of proposed antigens and reasonably considered mechanisms of antigen introduction during the surgical event

A false null hypothesis suggests that contaminants that caused DLK in these cases had to be introduced into the eye in a highly localized manner. We have no explanation of how this might occur, particularly since all cases were performed using a femtosecond laser. Perhaps worst of all, is that if we extrapolate this finding to all cases of Stage 4 DLK in which the major pathology also (apparently by definition[4-6]) occurs in the corneal center, then we must similarly conclude that contaminant introduction in Stage 4 DLK is focally directed, irrespective of the type of contaminant under consideration. This is difficult to reconcile with wide variety of proposed antigens and reasonably considered mechanisms of antigen introduction during the surgical event.

The alternative null hypothesis would then be that the contaminants in Stage 4 DLK are deposited during the surgical process in a highly focal manner. However, in this hypothesis, we have no mechanism to suggest that one location in the LASIK interface is more likely than another to become contaminated. In this case our null hypothesis is that the contaminant is introduced in a highly localized fashion but that the exact location of this focus would be randomly distributed in the interface. Table III illustrates the results of chi-square testing of this hypothesis again using the 8 eyes in this series. For simplicity, we divided the LASIK interface up into 5 arbitrary zones – center, superior, inferior, nasal and temporal and, for convenience, ignore the fact that an antigen could easily have multiple foci.

Table III - Chi-square analysis testing of the hypothesis that contaminants in these 8 cases were randomly distributed determines that this null hypothesis is false. Contaminants in these cases of Stage 4 DLK are not randomly distributed. They are apparently consistently and systematically deposited in the corneal center despite the absence of any known mechanism for how this might occur

Distribution of Pathology	Center	Superior	Inferior	Nasal	Temporal
Number Observed	8	0	0	0	0
Number Expected	1.6	1.6	1.6	1.6	1.6
P value	0.0000019				

Irrespective of how simple our algorithm and null hypothesis, once again, our chi-square analysis determines that the null hypothesis is false. Contaminants in these cases of Stage 4 DLK are not randomly distributed. They are apparently consistently and systematically deposited in the corneal center despite the absence of any known mechanism for how this might occur.

Another method for examining the second null hypothesis is to calculate the probability of finding, by chance, 8 consecutive eyes in which the major pathology happened to occur in the center of the cornea. If we again divide the cornea into the same five regions, the chance of any one of the five zones being affected (if the contaminant is randomly distributed) is approximately 20%. In other words, any given zone will be affected by a contaminant on average 2 times out of 10. The statistical odds of finding, by chance alone, eight eyes in a row in which the center of the cornea is exclusively affected is very small as it can occur in only one permutation. Specifically, the odds of finding such an event is 0.2^8 or 0.0000025. In essence, there are only 2.5 chances out of one million that such an event would occur by random chance alone.

The statistical odds of finding, by chance alone, eight eyes in a row in which the center of the cornea is exclusively affected is less than 2.5 in 1 million

Based upon statistical analysis, the notions that contaminants in DLK (or specifically Stage 4 DLK) are 1) distributed diffusely in the interface or; 2) they are distributed locally and randomly appear to be incorrect. However, the rejection of both null hypotheses would strongly suggest that contaminants in Stage 4 DLK are deposited in a local manner and the focus of their deposition is consistently the center of the cornea. Regardless of how unlikely such a mechanism might be, this latter idea actually appears to be consistent with ideas regarding disease pathogenesis frequently promulgated in the scientific literature.

> If chemotaxis is the mechanism behind Stage 4 DLK, in all known cases the highest concentration of the particular contaminant must be exclusively deposited in the corneal center. There is no known mechanism to accomplish such an extraordinary feat. Moreover, the odds of this happening by chance alone are likely worse than winning the state lottery. This is not a rational scientific position.

In light of this extraordinary level of disease focality for Stage 4 DLK, if chemotaxis is the mechanism behind Stage 4 DLK, we would need to postulate that in all known cases of Stage 4 DLK the highest concentration of the particular contaminant involved (out of the list of multiple antigens and toxins) is not just consistently, but exclusively deposited in the corneal center, irrespective of the clear absence of any known mechanism with which to accomplish such an extraordinary feat. Moreover, the odds of this happening by chance alone are likely worse than winning the state lottery. We do not believe this to be a rational scientific position.

We posit that such an improbable mechanism is unlikely to be the

> In our model, factors included on the list of etiologic agents ranging from endotoxins to sheet rock dust do cause DLK but do so by initiating a sequence of physiologic events that result in a defined pathologic process wherein the major pathology will predictably occur in the corneal center *entirely* because of the design and function of the cornea itself

physiologic basis for these disorders. In fact, our model rejects both of these null hypotheses and describes a pathophysiologic process in which multiple antigens or toxins initiate a final common disease pathway. In our model, factors included on the list of etiologic agents ranging from endotoxins to sheet rock dust do cause DLK but do so by initiating a sequence of physiologic events that result in a defined pathologic process wherein the major pathology will predictably occur in the corneal center *entirely* because of the design and function of the corneal tissue itself. As a consequence, our model is not based on highly improbable events or magical surgical processes.

> Inflammatory cells do not know where the center of the cornea is located any more than they know the location of the center of the universe

In many textbooks and published papers on DLK the mechanism by which inflammatory cells arrive at the center of the cornea is simply ignored[1-6,50]. We suspect that either the authors have no

> In the absence of such an unknown mechanism, we posit that any disease paradigm that invokes either antigen induced chemotaxis or "natural" cell migration as a core disease mechanism could reasonably be considered to be relying upon either extraordinarily improbable events or non-physiologic magical processes

idea why this process occurs or perhaps they assume that acute inflammatory cells simply possess the wherewithal to naturally navigate to the center of the cornea where they are proposed to aggregate, release lytic enzymes and cause matrix destruction. Unfortunately, unless acute inflammatory cells possess unique powers of navigation that are currently not known to exist, these cells simply have no inherent form of direction-finding or dead reckoning that would provide them with either a map, steering command sequences or directions to the center of the cornea. Based upon the lack of discussion of a mechanism in these authoritative sources, we suggest that many clinicians have fallen prey to the logical fallacy of anthropomorphistic thinking. Fundamentally, inflammatory cells in the eye do not know where the center of the cornea is located any more than they know the location of the center of the eye, the center of the body or the center of the universe.

As a consequence, unless we postulate a heretofore-unknown mechanism controlling cell migration in DLK, the location of virtually all tissue necrosis in the corneal center is highly problematic. In the absence of such an unknown mechanism, we posit that any disease paradigm that invokes either antigen induced chemotaxis or "natural" cell migration as a core disease mechanism could reasonably be considered to be relying upon either extraordinarily improbable events or non-physiologic magical processes.

> The location of primary pathology exclusively in the center of the cornea in CTK is equally challenging. To date, neither the identity, chemical structure nor the mechanism of production of such a "photoactivated toxin" has been identified

The location of primary pathology exclusively in the center of the cornea in CTK is equally challenging[22]. Certainly, if the inciting toxin is introduced at the time of surgery then its consistent location in the corneal center is highly unlikely. If we were to were to accept the position of Sonmez and Maloney that Stage 4 DLK is actually CTK[22], in our consecutive series of eight eyes, the probability of a toxin being place in the corneal center in each case is approximately 2.5 out of one million. In recognition of this defect in their theory, the authors postulate a heretofore unknown "phototoxin" derived from Betadine® produced as a result of excimer laser illumination is the cause. To be clear, in the series of eyes presented in this paper, none of the eyes were prepared using Betadine® or any other similar chemical. Moreover, to date, neither the identity, chemical structure nor the mechanism of production of such a "photoactivated toxin" has been identified. Apparently, the search for this illusive "photoactivated" toxic chemical remains ongoing.

However, despite the lack of viable mechanism or target molecule, the notion of a photoactivated or laser induced toxin is clearly attractive. Nevertheless, irrespective of the perceived convenience of this concept, we believe that this idea fails to retain credibility when examined logically.

Firstly, the mechanism of action of an excimer laser is that of photoablation. That is precisely why this technology is useful for etching silicone, plastic or corneal tissue. For an excimer laser to create a phototoxin the following events would need to occur. A contaminant, such as Betadine®, would need to be introduced onto the stromal bed at the precise moment that the flap is either created or manipulated, immediately prior to the photoablation process. In addition, the contaminant would need to be either diffusely distributed in the interface or, if localized, the contaminant would need to be located in the center of the cornea. Based on a knowledge of how LASIK surgery is typically performed, such a predictable type of contamination event would be highly unlikely to occur during routine LASIK surgery. Perhaps most disconcerting is the recognition that if and when this highly improbable contamination of the flap and interface

actually occurred, the action of the excimer laser beam will be to photoablate and physically remove any residual molecules of contamination from the stromal surface. As a consequence, rather than being transformed into some new chemical structure, the contaminant will be far more likely to be removed from the eye by the photoablation process.

> Despite the lack of viable mechanism or target molecule, the notion of a photoactivated or laser induced toxin is clearly attractive. Nevertheless, irrespective of the perceived convenience of this concept, we believe that this idea fails to retain credibility when examined logically

If the eyes in our series were suspected to be cases of CTK, the concept of a photoactivated toxin is equally dissatisfying. Two of the eyes were mixed astigmats and one demonstrated very high myopic astigmatism. Despite photoablation profiles that clearly were not isolated to the corneal center and included cases where the most laser pulses were placed peripherally, the major pathology in all three eyes was exclusively located and confined to the corneal center. Such a finding is simply not compatible with a photoactivated or excimer laser induced pathological paradigm.

Of additional concern is the assertion by Sonmez and Maloney[22] that Stage 4 DLK does not exist but rather has been mistakenly misidentified and in reality represents CTK and is unrelated to interface inflammation. This idea appears at odds with the spectrum of accounts both published and unpublished of severe interface inflammation leading to central "stromal melting" associated with entities ranging from endotoxins, aerosolized mold, skin marking pens or sheet rock dust. Apparently, in the photoactivation paradigm, this great diversity of materials is all conveniently converted to toxins by the excimer laser and these phototoxins then create essentially identical tissue responses in the cornea.

> Whether we contemplate the DLK or CTK disease paradigm, the location of virtually all tissue "necrosis" in the corneal center is highly problematic. Antigen induced chemotaxis or "natural" inflammatory cell migration as mechanisms require either a reliance on assigning human characteristics to cells (Logical Error of Anthropomorphism) or a dependence upon highly implausible events (Logical Error of Reliance on Highly Improbable Events)

In summary, irrespective of whether we contemplate the DLK or CTK disease paradigm, the location of virtually all tissue "necrosis" in the corneal center is highly problematic. We suggest that antigen induced chemotaxis or some form of "natural" inflammatory cell migration as pathologic mechanisms require either a reliance on assigning human characteristics to cells (Logical Error of Anthropomorphism) or a dependence upon highly implausible events (Logical

Error of Reliance on Highly Improbable Events).

We also believe that both the DLK and CTK paradigms suffer from the logical fallacy of 'Ignoring the Common Etiology'. If a patient presents to the doctor with a skin rash and a fever it would seem silly to suspect that the rash caused the fever or the fever caused the rash or consider the fever to be one disease and the rash another. The skin rash and fever are symptoms rather than causes of the disease. Although seemingly interrelated, both symptoms are likely the result of a viral infection. Unfortunately, this is the type of logic embraced by both the DLK and CTK paradigms. Our model will demonstrate that DLK and CTK, as well as the related entities of CFN, FNS and CLK, are caused by a sequence of physiologic events that result in a defined pathologic process wherein the major pathology and symptoms predictably occur in the corneal center entirely because of the design and function of the corneal tissue itself. Our model describes the mechanism of this common etiology.

> Our model will demonstrate that DLK and CTK, as well as the related entities of CFN, FNS and CLK, are caused by a sequence of physiologic events that result in a defined pathologic process wherein the major pathology and symptoms predictably occur in the corneal center entirely because of the design and function of the corneal tissue itself

What causes the frequently observed hyperopic shift?

For Stage 4 DLK, CTK, CFN, FNS and CLK, the hyperopic shift has been generally attributed to tissue necrosis in the visual axis[1-6,9,22,24-26,50]. Despite the general acceptance of this belief, a thorough review of the literature reveals no objective scientific data to support this position. Irrespective of the latter, this paradigm is frequently reiterated in textbooks on LASIK[5,6,50] as well as being vigorously and often dogmatically defended in scientific symposia on the topic (BRW, personal observation).

> The hyperopic shift has been generally attributed to tissue necrosis in the visual axis. Despite the general acceptance of this belief, a thorough review of the literature reveals no objective scientific data to support this position

However, the current study demonstrates that topical hyperosmotics can very rapidly reverse hyperopic shifts in Stage 4 DLK and that patients can consistently completely recover without permanent central corneal scarring, residual refractive error or other adverse optical effects. Numerous other observers have reported that corneal healing over time can frequently reverse the hyperopic shift currently attributed to tissue necrosis for both Stage 4 DLK and CTK[22,52-54].

> This study demonstrates that topical hyperosmotics can very rapidly reverse hyperopic shifts in Stage 4 DLK and that patients can consistently completely recover without permanent central corneal scarring, residual refractive error or other adverse optical effects. Numerous other observers have reported that corneal healing over time can frequently reverse the hyperopic shift currently attributed to tissue necrosis for both Stage 4 DLK and CTK

As a result, contrary to previous views regarding the adverse impact of Stage 4 DLK and CTK on vision where vision loss from central necrosis and scarring lead to inevitable loss of BCVA[1-6,50], it seems that we are now to reverse direction and believe that liquefaction necrosis in the central visual axis is frequently a self-limited disease. Moreover, other than to suggest that other equally undocumented sources of healing such as epithelial hyperplasia may be a factor, the basis for how or why some cases of Stage 4 DLK can recover without any adverse visual sequelae has essentially been left moot by

> Other than to suggest that undocumented sources of healing such as epithelial hyperplasia may be a factor, the basis for how or why many cases of Stage 4 DLK can recover without any adverse visual sequelae has essentially been left moot by the Ophthalmic profession

the profession. This is particularly vexing since complete recovery of tissue form or function is not typical for corneal healing post tissue necrosis from infectious keratitis or other corneal inflammatory disorders where permanent stromal scarring and loss of BCVA is the norm. It is also completely atypical for healing following LASIK. Cases of decentered ablations or irregular astigmatism rarely spontaneously improve over time. In our view, it is unacceptable to suggest or believe that epithelial hyperplasia can consistently create an optimal visual outcome for Stage 4 DLK when decades of clinical observations have demonstrated that the process of epithelial remodeling is nearly always unable to restore vision in other necrotic processes in the cornea or to restore gross lesions involving loss of tissue volume in the body in general.

> In our view, it is unacceptable to suggest or believe that epithelial hyperplasia can consistently create an optimal visual outcome for Stage 4 DLK when decades of clinical observations have demonstrated that the process of epithelial remodeling is nearly always unable to restore vision in other necrotic processes in the cornea

Sonmez and Maloney have further suggested that instead of central matrix necrosis in CTK, the proposed loss of corneal tissue volume causing macrostriae and hyperopia is related to keratocyte apoptosis and that this causes the cornea to collapse upon itself[22]. Collapse of ECM tissue resulting from cell loss alone

has, to our knowledge, never been previously observed in any tissue, this concept has not been validated by confocal microscopy or other means of microscopic tissue examination and, perhaps most importantly, this mechanism is not compatible with known physiologic behaviors of interstitial cells or the surrounding extracellular matrix following cell apoptosis. On the contrary, a broad spectrum of scientific literature describes a propensity for the ECM to expand with the loss of interstitial cell function[55]. As a consequence, the notion that cell mediated tissue collapse is an explanation for a hyperopic shift is completely at odds with currently understood principles of tissue physiology and the active role of interstitial cells in maintaining tissue

> Sonmez and Maloney suggest that instead of central matrix necrosis in CTK, the proposed loss of corneal tissue volume is related to keratocyte apoptosis. The notion of tissue collapse from apoptosis as an explanation for a hyperopic shift is completely at odds with currently understood principles of tissue physiology and the active role of interstitial cells in maintaining tissue volume and fluid dynamics. Keratocyte apoptosis as a mechanism that produces volume loss in the central cornea is simply not supported by any scientific data

volume and fluid dynamics. Keratocyte apoptosis as a mechanism that produces volume loss in the central cornea is simply not supported by scientific data and the promotion of such an idea underscores the degree of magical thinking and general failure to base theories on credible scientific research that is currently prevalent in discussions of these disorders.

> If hyperopic shift in Stage 4 DLK, CTK, CFN, FNS and CLK is caused by tissue or cell necrosis in the central cornea, the frequent clinical observation of complete reversibility of the effect is highly problematic as no physiologic mechanism of healing is currently known to exist that can consistently reverse tissue necrosis without creating either scarring or residual refractive anomalies

In summary, if hyperopic shift in Stage 4 DLK, CTK, CFN, FNS and CLK is caused by tissue or cell necrosis in the central cornea, the frequent clinical observation of complete reversibility of the effect is highly problematic as no physiologic mechanism of healing is currently known to exist that can consistently reverse tissue necrosis without creating either scarring or residual refractive anomalies. If such a novel healing mechanism does exist, we suggest that it demands immediate study as its potential application to other necrotic disorders of the cornea could be extraordinarily valuable in the management of such diseases. However, in the absence of such a mechanism, we express serious concern over what we perceive as a disturbing trend towards proposing non-physiologic explanations in an attempt to explain such clinical dilemmas.

Clinical intra-operative examination findings in this study are not consistent with tissue necrosis

As noted in the clinical case descriptions of this series of eyes, those cases that underwent flap lift combined with rinse or stretch procedures did not show any evidence of stromal tissue necrosis, abnormal tissue friability in the stromal bed or thinning or weakening of the flap, irrespective of the fact that all of these Intralase™ created flaps were very thin and all intraoperative observations were made by a highly experienced refractive surgeon (BRW). This observation was unexpected since preoperatively the posterior stroma appeared to have significant thinning in the area of central opacification and the flap and anterior stroma appeared necrotic and white under slit lamp examination. We did observe intraoperatively however that the central epithelium in these cases was very loose

> No eyes in this series showed any evidence of stromal tissue necrosis, abnormal tissue friability in the stromal bed or thinning or weakening of the flap, irrespective of the fact that all of these Intralase™ created flaps were very thin and all intraoperative observations were made by a highly experienced refractive surgeon

and edematous. A third observation was that immediately upon completion of a flap lift and irrigation procedure, slit lamp examination demonstrated the absence of inflammatory cells in the interface, however, the central corneal opacification was unchanged from its preoperative condition indicating that the central whitening was not due to inflammatory cell clumping in the interface as has been commonly suggested.

All of these observations are inconsistent with clinical narratives published by other authors and in textbooks on LASIK surgery[1-6,9,22,24-26,50]. This is of significant concern. However, we believe that this concern must be tempered by the observation that these previously published observations are entirely subjective in nature. As such, they are

> In the current scientific literature there is no intraoperative video or other photographic clinical evidence included in any of these papers to definitively support the notion that necrosis of the residual stromal bed or LASIK flap actually occurs

subject to the preconceived notions and surgical experience of the surgeon involved. Most importantly, we must acknowledge the fact that in the current scientific literature there is no intraoperative video or other photographic clinical evidence included in any of these papers to definitively support the notion that necrosis of the residual stromal bed or LASIK flap actually occurs. This lack of actual evidence simply adds to the controversy and confusion.

In addition to the lack of scientific data to support various clinical observations,

there is also no consensus amongst authoritative sources with respect to the location of pathology in DLK. To illustrate the apparent confusion among authors on this topic, we note that Dr. Machat in his textbook on LASIK

> In addition to the lack of scientific data to support various clinical observations, there is also no consensus amongst authoritative sources with respect to the location of pathology in DLK

surgery[5] states that in the most severe Stage 3 DLK (in the Machat classification Stage 3 is equivalent to the Linebarger Stage 4[4,50]) "the infiltrate is limited to the interface, and although the bed may be minimally involved, the undersurface of the flap appears uninvaded." In the Smith and Maloney[1] paper they state that "the infiltrate has the following characteristics: It is confined to the interface, extending neither anteriorly into the flap nor posteriorly into the stroma". Of additional interest is that a careful examination of Linebarger et al.[4] and two other textbooks on LASIK[6,50] reveals the fact that they simply do not state whether the 'stromal melt' occurs in the interface, flap or residual stromal bed and essentially gloss over any distinction between the three areas. This is in stark contrast to the belief of Jonathan Christenbury MD, a highly experienced refractive surgeon. Dr. Christenbury believes that in Stage 4 DLK the necrosis is completely isolated to the flap and advocates complete excision of the flap to allow the cornea to heal. He has described this treatment in detail on the Intralaser bulletin board online and at the Intralase™ Users Meeting (Chicago, April 2008). Although not published in a peer reviewed paper, he reports that such treatment results in excellent clinical outcomes (personal communication). Similar to Dr. Christenbury, Hainline et al. [24] in their description of Central Flap Necrosis determined that the stromal bed was normal but the flap was deemed "necrotic".

Even if one accepts the position that each of these observations are accurate, the concept of a mechanism wherein a necrotic event can be isolated to a single area of the cornea, in the absence of any known barrier to extension of that process, is simply ignored. These authors and the scientific literature as a whole is surprisingly moot on the question of how any known necrotic processes can be isolated to either the flap, interface or bed.

> A mechanism wherein a necrotic event can be isolated to a single area of the cornea, in the absence of any known barrier to extension of that process, is simply ignored. These authors and the scientific literature as a whole is surprisingly moot on the question of how any known necrotic processes can be isolated to either the flap, interface or bed

Some insight into the magical thinking involved in such discussions regarding location of the necrosis can be obtained by a detailed review of the paper by Hainline et al.[24] on Central Flap Necrosis. In their series of eyes exhibiting Central Flap Necrosis they did not note any abnormality in the stromal bed intraoperatively. However, the authors did describe a "jelly like" consistency to the central flap, which they considered to represent tissue necrosis localized to the flap. This paper demonstrates many of the logical inconsistencies that appear to be prevalent in discussions regarding DLK and related entities. Specifically;

1. All patients with this marked central flap necrosis recovered good visual acuities without adverse visual aberrations. No known mechanism exists to explain this remarkable recovery in this series of patients and no mechanism is postulated by the authors.

2. No logical explanation was presented to suggest why or how a necrotic process caused by any known toxin or factor introduced during LASIK surgery can become exclusively and consistently isolated to the central flap, regardless of the clear absence of any barrier to extension of that necrotic process to surrounding tissue, including the stromal bed.

3. No tangible evidence to support tissue necrosis is provided other than a description of subjective intraoperative or slit lamp observations

4. The mechanism proposed for the cause of the necrosis (Cidex® antiseptic, Civco, Kalona, Iowa, USA) was clearly refuted by their own clinical data wherein a significant percentage of cases did not receive Cidex®

> In our view, the idea that Central Flap Necrosis represents a reversible form of tissue necrosis that is exclusively confined to the flap represents magical thinking and is challenging to explain using normal physiologic principles

Irrespective of the proposed etiologies for flap necrosis expressed by Hainline et al. [24], the notion that a necrotic process can consistently remain isolated to the flap thereby causing no adverse effect on the adjacent residual stromal bed, in the absence of any known barrier to extension of such a process, is challenging to explain using normal physiologic principles. Moreover, the authors failed to provide any tangible photographic evidence to support their intraoperative observations. In addition, the only factor actually identified and proposed as a cause for the flap necrosis was residual Cidex® on the tonometer used in association with the microkeratome. No explanation as to how or why Cidex® toxicity could be reasonably expected to be confined to the flap was offered. No explanation was provided for why this same flap necrosis occurred in the absence of Cidex® and the tonometer. Neither was any explanation provide for why only select patients were affected even when the tonometer was used on other patients in the same surgical setting or why these other patients showed no apparent ill effects such as epithelial injury or even minor amounts of epithelial or stromal edema. The proposed mechanism is simply not consistent with their data.

In our view, the idea that Central Flap Necrosis represents a reversible form of tissue necrosis that is exclusively confined to the flap represents magical thinking. Although, as described in greater detail in this paper, we recognize that our model does anticipate that the flap and residual stromal bed do exhibit different performance characteristics and will behave independently. Our model will also explain why the flap opacity observed by Hainline et al.[24] was cone shaped, why the opacity occurred with and without exposure to Cidex® and why the opacity occurred in Femtosecond laser cases as well as microkeratome surgery. We also believe our model to be based on sound basic and clinical science research and in agreement with known physiologic principles.

Based upon our experience, we do agree with the authors that the stromal bed in these cases was normal. However, we postulate that the "jelly like" consistency of the central flap noted by Hainline et al. [24] in CFN was likely produced by stromal edema combined with loose edematous epithelium centrally, rather than stromal

> We find it distinctly refreshing to find authors that understand that the stromal bed is normal in Stage 4 DLK and related entities. This observation does serve to reinforce the notion that our understanding of truth is progressive and our grasp of that truth often incremental. Recognizing that the residual stromal bed is normal in these disorders is a monumental step forward

necrosis. We also agree with the observation by Dr. Christenbury that the stromal bed is normal in cases of Stage 4 DLK. In fact, we find it distinctly refreshing to find authors that understand that the stromal bed is normal in Stage 4 DLK and related entities. This observation does serve to reinforce the notion that our understanding of truth is progressive and our grasp of that truth often incremental. Recognizing that the residual stromal bed is normal in these disorders is a monumental step forward. We posit that realizing that "necrosis" of the flap is not actually necrosis but is focal tissue edema that only appears to be necrosis due to the intense light scattering produced by the slit lamp merely represents the next stage of understanding of these disorders. It is also a concept that is apparently already supported by Dr. Machat[5].

> In the series of patients in this paper, intraoperative observations of the LASIK flap and residual stromal bed in Stage 4 DLK by a highly experienced refractive surgeon (BRW) are not consistent with stromal necrosis of either the flap or bed

In the series of patients in this paper, intraoperative observations of the LASIK flap and residual stromal bed in Stage 4 DLK by a highly experienced refractive surgeon (BRW) are not consistent with stromal necrosis. Moreover, in current scientific literature there is no intraoperative video or other photographic clinical evidence to support the notion that necrosis of the residual stromal bed or

LASIK flap actually does occur. In addition, as described in greater detail in other sections this paper, the idea that tissue necrosis is a cause for a hyperopic shift is also not supported by histopathology. Unfortunately, at present, the primary clinical evidence for tissue necrosis in Stage 4 DLK, CTK, CFN, FNS and CLK appears to be entirely presumptive in nature and is based upon the observation of a hyperopic shift and central flap macrostriae[1-6,9,22,24-26,50]. Clearly, if an alternative explanation for a reversible hyperopic shift can be identified, the primary clinical evidence used to support the notion of stromal necrosis would no longer exist.

> Realizing that "necrosis" of the flap is not actually necrosis but is focal tissue edema that only appears to be necrosis due to the intense light scattering produced by the slit lamp merely represents the next stage of understanding of these disorders. It is also a concept that is apparently already supported by Dr. Machat

Why is there no evidence for tissue necrosis in Stage 1, 2 or 3 DLK?

Tissue necrosis in Stage 4 DLK is believed to occur as a consequence of the release of lytic enzymes by inflammatory cells or the activation of matrix metalloproteinases (MMP) secreted by these inflammatory cells[1-6,9,50]. Based upon the notion that tissue matrix necrosis occurs as a result of these chemical factors, it is difficult to rationalize the apparent lack of 'continuum of effect' of such processes as is currently conceived in the pathogenesis of DLK[23].

Specifically, it is reasonable to anticipate that MMP's and lytic enzymes released by acute inflammatory cells in Stage 1, 2 or 3 would demonstrate some adverse clinical effect. Lytic enzymes and MMP's, even if released by inflammatory cells in relatively small amounts, should create some clinically detectable amounts of

In the current disease paradigm for DLK, tissue necrosis apparently occurs only when the concentration of MMP's and lytic enzymes reaches an unidentified threshold value wherein massive stromal matrix destruction occurs and this threshold occurs exclusively in the center of the cornea rather than in an incremental continuum. This is nonsense and "non-science"

necrosis in the peripheral stroma. There is at present, no rational proposal that would explain why such a release or effect

should not occur, particularly when peripheral infiltrates in Stage 1-3 can be intense[6]. Although such necrosis may be less than that observed in Stage 4 DLK, some necrosis mediated by lytic enzymes would likely occur in such circumstances in some cases nevertheless. As a result, one should predict that the spectrum of pathology in DLK should range between very small amounts of peripheral stromal necrosis with mild disruption on topography or wavefront measurement and perhaps with associated glare, halos and mild vision distortion to massive amounts of tissue loss across the entire breadth of the cornea combined with profound vision impairment. Moreover, the location(s) of such tissue necrosis would be anticipated to occur randomly throughout the LASIK interface or perhaps even exhibit multiple disease foci.

The lack of an observable continuum of effect and transition from mild necrosis to severe focal necrosis is troubling, particularly when Stages 1, 2 or 3 DLK are observed to demonstrate progressive intensities of peripheral interface and stromal inflammation

However, in the current disease paradigm for DLK, tissue necrosis apparently occurs only when the concentration of MMP's and lytic enzymes reaches an unidentified threshold value wherein massive stromal matrix destruction occurs and this threshold occurs exclusively in the center of the cornea rather than in an incremental continuum. As a result, based upon current disease theory, measurable pathology in DLK only occurs in a highly peculiar 'binary' fashion where the disease presentation is always a 'zero' or a 'one'.

The lack of an observable continuum of effect and transition from mild necrosis to severe focal necrosis is troubling, particularly when Stages 1, 2 or 3 DLK are observed to demonstrate progressive intensities of peripheral interface and stromal inflammation. Similar concerns can be expressed regarding the apparent lack of transitional forms of CTK as well. However, as a result, in clinical practice there are actually only two relevant clinical entities in DLK. Stage 1 through 3 are essentially identical as they all have the same excellent outcome and any distinction appears clinically irrelevant, while Stage 4 DLK results in permanent scarring and tissue loss isolated to the corneal center. We suggest that such a clinical staging mechanism is flawed, dysfunctional and illogical.

Of additional concern is that MMP's are generally secreted by live and active inflammatory cells rather than during cell decay[56]. Moreover, these enzymes are typically secreted as pro-enzymes and require cleavage and activation before they are effective with the latter being usually accomplished by bacterial proteases. Further compounding the confusion is that, unfortunately, despite the general acceptance of this pathophysiologic mechanism for tissue destruction in Stage 4 DLK as near dogma[5,6,50], to date, no study has demonstrated let alone quantified the amount or type of matrix metalloproteinases and collagenases present within the interface.

What causes the pathognomonic "shifting sands" appearance in DLK?

No current theories have either addressed or accounted for the pathognomonic shifting sands appearance of DLK (R Maddox MD, A Hatsis MD. "Sands of the Sahara" Poster: Symposium on Cataract, IOL and Refractive Surgery, San Diego, CA, USA, April 1998 and JL Gunn, SL Forstot, A Hatsis et al. "Sands of the Sahara: Post LASIK Interface Inflammation – Reality of Mirage?" Poster: Symposium on Cataract, IOL and Refractive Surgery, San Diego, CA, USA, April 1998). Apparently, the value in identifying the mechanism that creates this characteristic feature of DLK has not been viewed as an important component to a comprehensive pathophysiologic description of this disorder.

> Identifying the mechanism that creates the pathognomonic shifting sands characteristic of DLK has not been viewed as an important component to a comprehensive pathophysiologic description of this disorder

Why do Stage 4 DLK and CTK appear to be self-limited disorders?

A number of studies support the position that patients affected by either disorder may recover full visual function over an extended period of time[22,52-54]. In the consecutive series reported in this study of patients with Stage 4 DLK, that timeline to complete recovery ranged from 12 weeks to 9 months. Speed of recovery was accelerated in later cases when a more focused approach was taken to reversing tissue edema.

In the 10 years since the first description of DLK there have been no published reports of corneal transplant data from a single case of Stage 4 DLK. We find the scientific literature's silence on the topic of reports of penetrating or lamellar keratoplasty in such a large pool of patients to be deafening

In addition, it is surprising to note that in the 10 years since the first description of DLK by Dr. Robert Maddox (R Maddox MD, A Hatsis MD. "Sands of the Sahara" Poster: Symposium on Cataract, IOL and Refractive Surgery, San Diego, CA, USA, April 1998), there have been no published reports of corneal transplant data from a single case of Stage 4 DLK. Published reports suggest that Stage 4 DLK occurs at a rate of 1 in 5,000 eyes[4]. With over 10 million LASIK cases performed, the worldwide pool of Stage 4 DLK patients can be reasonably estimated to be

In the 8 eyes in this series, 100% recovered BSCVA of 20/20 and exhibited no complaints of adverse optical effects such as glare or halos. If Stage 4 DLK or CTK are caused by central corneal necrosis, such a series is truly extraordinary

somewhere around 2,000 eyes. We find the scientific literature's silence on the topic of reports of penetrating or lamellar keratoplasty in such a large pool of patients to be deafening.

In the 8 eyes in this series, 100% recovered BSCVA of 20/20 and exhibited no complaints of adverse optical effects such as glare or halos. If Stage 4 DLK or CTK are caused by central corneal necrosis, such a series is truly extraordinary. To illustrate how improbable such an event might be in clinical practice we have included a chi-square analysis of potential visual outcomes based upon assumptions of probability of obtaining a full visual recovery (Table IV).

Table IV - In the 8 eyes in this series, 100% recovered BSCVA of 20/20 and exhibited no complaints of adverse optical effects such as glare or halos. If Stage 4 DLK or CTK are caused by central corneal necrosis, such a series is truly extraordinary. To illustrate how improbable such an event might be in clinical practice we have included a chi-square analysis of potential visual outcomes based upon assumptions of probability of obtaining a full visual recovery. Even if we make the seemingly outrageous assumption that an eye affected by central necrosis of the flap and anterior residual stromal bed in Stage 4 DLK can recover full visual function with 20/20 BSCVA 50% of the time, our data disprove that hypothesis. The odds of finding a series of 8 eyes that all recover a BSCVA of 20/20 by chance alone, if the probability of healing to that level occurs 5% of the time, is 0.05^8 or 3.0×10^{-11}. If the odds of full corneal healing are improved to 25%, the likelihood of finding such an 8 eye series improves to only 0.25^8 or 1.5×10^{-5} or 15 times out of a million

Final BSCVA	20/20 or Better	Worse than 20/20	P value
Number Observed	8	0	
Number Expected if 5% recover BSCVA of 20/20	0.4	7.6	6.33 X 10-35
Number Expected if 20% recover BSCVA of 20/20	1.6	6.4	1.54 X 10-8
Number Expected if 50% recover BSCVA of 20/20	4	4	0.0046

In a series of 8 eyes, even if we make the seemingly outrageous assumption that an eye affected by central necrosis of the flap and anterior residual stromal bed in Stage 4 DLK can recover full visual function with 20/20 BSCVA 50% of the time, our data disprove that hypothesis. The odds of finding a series of 8 eyes that all recover a BSCVA of 20/20 by chance alone, if the probability of healing to that level occurs 5% of the time, is 0.05^8 or 3.0×10^{-11}. If the odds of full corneal healing are improved to 25%, the likelihood of finding such an 8 eye series improves to only 0.25^8 or 1.5×10^{-5} or 15 times out of a million.

> We posit that the reason that these eyes all recovered is because the pathology located in the center of the cornea is simply not necrosis

It is our position that we are not exceptionally lucky nor that we have discovered a novel treatment regimen that stimulates pristine healing of keratocytes, the ECM, collagen fibrils and epithelium or that corneal tissue exhibits some novel form of regeneration when exposed to an excimer laser beam, antigens and toxins or even that corneal epithelium can consistently achieve extraordinary levels of performance in acting to uniformly fill irregular "calderas" in the stroma. Rather, we posit that the reason that these eyes all recovered is because the pathology located in the center of the cornea is simply not necrosis.

> Current paradigms regarding disease pathology in Stage 4 DLK, CTK, CFN, FNS and CLK propose that liquefaction necrosis centered in the visual axis can frequently resolve without residual scarring or measurable visual compromise

Despite these findings, current paradigms regarding disease pathology in Stage 4 DLK, CTK, CFN, FNS and CLK appear to propose that liquefaction necrosis

centered in the visual axis can frequently resolve without residual scarring or measurable visual compromise[22,52-54]. Such an observation is particularly curious when even mildly decentered laser photoablations resulting from PRK or LASIK are highly unlikely to spontaneously improve with time. Moreover, corneal healing following tissue necrosis from microbial or infectious keratitis, autoimmune processes or lamellar surgery rarely, if ever, appear to spontaneously heal in this manner.

Sonmez and Maloney[22] suggested that the loss of corneal tissue volume in CTK is caused by cell loss alone without loss of the extracellular matrix (ECM), allowing the cornea to collapse upon itself. Unfortunately, such a proposal lacks scientific support as it is not corroborated by histological data and is at odds with a broad base of basic science research which indicates that the typical biological responses of the ECM is to expand with interstitial cell injury[55].

> Corneal healing following tissue necrosis from microbial or infectious keratitis, autoimmune processes or lamellar surgery rarely, if ever, appear to spontaneously heal in this manner

The self-limiting characteristic of Stage 4 DLK and CTK is highly problematic for either the tissue or cell necrosis paradigm. As a consequence, in an attempt to account for this disease feature and their particular disease paradigm, it appears that many authors have found it convenient to propose mechanisms for corneal healing that currently fall outside the bounds of known science.

We posit that such thinking involves an appeal to logical fallacy of the "Highly Improbable Hypothesis". We suggest that prior to offering a bizarre or far-fetched hypothesis as the correct explanation, it is critical to first rule out more mundane explanations. The well worn medical idiom that suggests that "when one hears hoof beats in Texas, one shouldn't suspect Zebras" articulates this perspective. If we were to observe a field with a mutilated cow and flattened grass, we should not presume that aliens have landed in a flying saucer and savaged the cow to learn more about the beings on our planet. Similarly, in understanding complex problems in the cornea, we should first examine all theories that reasonably rely upon scientific research as to causation prior to postulating magical or highly speculative mechanisms with little or no scientific basis.

> In understanding complex problems in the cornea, we should first examine all theories that reasonably rely upon scientific research as to causation prior to postulating magical or highly speculative mechanisms with little or no scientific basis. If we were to observe a field with a mutilated cow and flattened grass, we should not presume that aliens have landed in a flying saucer and savaged the cow to learn more about the beings on our planet

Current Classification System is not a Clinically Relevant Tool

Clinical classification systems typically organize disease entities based upon common characteristics, etiologies or pathophysiologic pathways. Unfortunately, in the case of DLK, Stage 4 DLK, CTK, CFN, FNS and CLK current methods of organization frequently identify features that are symptoms of the disease as factors that cause the disorder, they generally rely upon speculation regarding etiology rather than specific knowledge and they fail to provide a final common pathophysiologic pathway that ties the entities together.

> Affixing the terminology of "multifactorial" to this disease complex appears to only apply a coating of gloss directed towards the goal of assigning a scientific hue and quality to a classification scheme more appropriately viewed as organized chaos

In general, causation for these disorders is considered to be multi-factorial[1-6,9-12,22,24-26,50,57]. This label suggests that multiple factors cause a final disease endpoint. Unfortunately, DLK, Stage 4 DLK, CTK, CFN, FNS and CLK are not truly multifactorial in origin. As a group they demonstrate highly similar clinical presentations that are currently classified as distinctly unique entities having no common etiology and no common pathophysiologic pathway. In our perspective, affixing the terminology of "multifactorial" to this disease complex appears to only apply a coating of gloss directed towards the goal of assigning a scientific hue and quality to a classification scheme more appropriately viewed as organized chaos.

Such a system frequently confounds disease prevention. Clusters of affected patients often appear to occur with no obvious etiology[12,33,51]. In other situations, known etiologic factors that are clearly present in the operative milieu do not appear to cause the disease[12,29,33,51,58,59]. The lack of a consistently identifiable cause makes preventing these disorders particularly vexing. Best management practices for treatment are equally difficult to develop. Without a clear understanding of the etiology of the disease, treatment methodologies are currently based entirely on empirically derived management procedures rather than targeted therapies proven effective using rigorous clinical trials[4-6,12]. As a result, it is our position that medical practice in these disorders is frequently driven by a pervasive form of anecdotally derived, ill-conceived dogma supported by only scant scientific logic or factual information.

> Without a clear understanding of the etiology of the disease, treatment methodologies are currently based entirely on empirically derived management procedures rather than targeted therapies proven effective using rigorous clinical trials

The scientific literature suggests that several types of DLK exist[4-6,50]. This classification system is based upon relative intangibles such as the frequency of occurrence of the disorder[8,12,13,46], the timing of disease onset or even the amount of inflammatory cells subjectively observed at random times by a clinician[22]. As a result, the classification system consists of four stages of DLK, combined with four distinctly different etiologies for DLK as well as CTK, CFN, FNS and CLK. Despite the fact that the final disease entities are clinically indistinguishable, there is no common final pathway or common clinical basis for these various "diseases".

> The classification system consists of four stages of DLK, combined with four distinctly different etiologies for DLK as well as CTK, CFN, FNS and CLK. Despite the fact that the final disease entities are clinically indistinguishable, there is no common final pathway or common clinical basis for these various "diseases"

One type of DLK occurs in "epidemics" while another appears to occur only "sporadically"[8,12,13,46]. Irrespective of the fact that these terms remain undefined and highly subjective, the frequency of presentation thereby creates a basis for classification. In general, epidemic DLK is believed to occur as a consequence of contamination of the surgical process by antigens such as lipopolysaccharide residues or biofilms[6,57], while sporadic DLK may be more associated with epithelial trauma[40,58,60-69] or atopy[45-47] and the peri-operative release of inflammatory cytokines[7,14-21]. In clinical practice, no actual data exists that can support this classification scheme as it is based almost entirely upon speculation on a case-by-case basis rather than actual scientific data gathered by the surgeon.

> In clinical practice, no actual data exists that can support this classification scheme as it is based almost entirely upon speculation on a case-by-case basis rather than actual scientific data gathered by the surgeon

Perhaps most perplexing to explain in the context of the current antigen deposition and acute inflammation paradigm is the occurrence of late onset DLK[61,62,70-78]. Cases of interface inflammation have been observed weeks and months after uncomplicated LASIK procedures that exhibited no DLK at time of the primary procedure. If the inciting factor for acute inflammation in the interface is an antigen or toxin deposited at the time of the initial LASIK surgery, then the lack of initial immune or tissue response coupled with a marked late response is troubling.

Many cases of DLK appear to defy such categorization. Concepts of antigen deposition such as biofilm contaminants appear unhelpful in explaining epidemics of DLK cases that occur in Femtosecond laser cases[43,79-83] employing dry heat sterilizers, single use sterile vacuum apparatus and new surgical instruments (Will BR. Diffuse Lamellar Keratitis and the IntraLASIK™ Procedure. Intralase™ Users Meeting, Dana Point, California,

June 2002) or the observation that rates of DLK decrease with optimization of femtosecond laser energy[43,79] (Will BR. Continuous Plane Photodisruption – Creating the Ideal Flap using Optimized Laser Parameters. Intralase™ Users Meeting. New Orleans, Louisiana. October 2004). Such observations have caused some femtosecond laser users to consider such interface inflammation occurring frequently or infrequently to be "pseudo-DLK". Dr. Perry Binder appears to have coined the novel term "focal lamellar keratitis (FLK)" for femtosecond laser related interface inflammation, which he suggests differs from DLK (P Binder MD. Femtosecond Laser Flaps: Managing Complications. Review of Ophthalmology. Jobson Publishing. February 2008). Whether Dr. Binder believes FLK to be an entirely new disease entity is currently unclear. If so, we may need to add this new acronym to

> Many cases of DLK appear to defy such categorization. The currently accepted complex and convoluted disease classification for DLK appears to declare that within the four stages of DLK there are at least four uniquely different disease variants including: 1) epidemic; 2) sporadic; 3) late onset and; 4) unclassifiable "pseudo-DLK" or "FLK" all of which are believed to demonstrate fundamentally different etiologies.

the expanding list of unique diseases. However, Dr. Binder has further suggested that the Femtosecond laser does not cause DLK but rather creates peripheral interface inflammation (Public statements in role as Medical Director and moderator at Intralase™ Users Meeting, Chicago, USA April, 2008 *and* Dalton M. A Worrisome Complication of LASIK. EyeWorld. June 2008). Provided the general belief that DLK is predominantly a multifactorial interface inflammatory disease, such a distinction is puzzling.

The currently accepted complex and convoluted disease classification for DLK appears to declare that within the four stages of DLK there are at least four uniquely different disease variants including: 1) epidemic; 2) sporadic; 3) late onset and; 4) unclassifiable "pseudo-DLK" or "FLK" all of which are believed to demonstrate fundamentally different etiologies. Antigen deposition in the LASIK interface that induces inflammatory cells to race to the region driven by chemotaxis is fundamentally a different disease mechanism than non-specific inflammation caused by cytokine release from damaged corneal epithelium.

> Rather than simplifying disease classification and management protocols, the introduction of CTK as a novel disease paradigm appears to only add an additional layer of complexity and confusion

Rather than simplifying disease classification and management protocols, the introduction of CTK by Sonmez and Maloney[22] as a novel disease paradigm appears to only add an additional layer of complexity and confusion. Although Sonmez and Maloney[22] propose that CTK is a distinctly different disorder with a distinctly different etiology from DLK, they admit that there is a very high frequency of association between DLK and CTK. In fact, they suggest that "stage IV lamellar keratitis"…shares no clinical features with diffuse lamellar keratitis except for its frequent coexistence."[22]. Instead of one disease entity, the authors apparently believe that these eyes demonstrate two independent disease processes simultaneously and that neither disease has a common cause or disease mechanism. We find this logic problematic and suggest that this paradigm demonstrates the logical fallacy of "failing to appreciate a common cause".

> Instead of one disease entity, CTK eyes "frequently" demonstrate two independent disease processes simultaneously and neither disease has a common cause or disease mechanism. We find this logic problematic and suggest that this paradigm demonstrates the logical fallacy of "failing to appreciate a common cause"

Unfortunately, instead of applying a healing salve to an apparently defective classification scheme, the notion of CTK and DLK existing as distinct and separate entities appears to have further confounded the scientific literature and our profession. An excellent example of this is the paper by Hadden et al.[32] chronicling an outbreak of DLK associated with use of a gentian violet marking pen. We agree with the author's premise that skin marking pens are a cause of DLK as we had a similar experience with gentian violet dye in 2001. Our experience ultimately led to our formal recommendation to Alcon Laboratories (Fort Worth, Texas) advising them to select a marking pen that utilized Crill dye combined with amylacetate and 2-methoxyethanol as solvents as their pen of choice for marking the eye for cyclotorsion control on the LADARVision® platform. Our concern however, is the author's classification of those patients that quickly recovered full visual function as exhibiting Stage 1 through 3 DLK while their worst DLK cases with "dense centrally demarcated... keratolysis", hyperopic shift and "marked central corneal thinning" they classified as having CTK. This description and classification appears completely at odds with the scientific literature. Moreover, the title of the paper "Outbreak of diffuse lamellar keratitis caused by marking-pen toxicity" and the author's conclusions that "marking pens or their associated packaging were the most likely cause of the outbreak of DLK" with no mention whatsoever of CTK, simply appear to blur any distinction in the author's minds between DLK and CTK other than a preoccupation with applying labels to patient presentations. Specifically, the author's appear to both assume and conclude that:

1. Gentian violet marking pens cause DLK and DLK subsequently causes CTK or;
2. Gentian violet marking pens cause both DLK and CTK

Irrespective of which conclusion is selected, these are both novel assertions that currently are not supported by published literature in either the DLK or CTK disease paradigms. Despite these conclusions, no explanation is offered that would provide the reader with any idea of how either disease cascade might occur independently from or conjointly with the other or how ink or solvents might concomitantly behave as both an antigen and a toxin. Further, the notion of how either gentian violet dye, its solvent or some component of the sterile packaging could act as a photoactivated toxin appears to be completely ignored.

> Provided the anecdotal nature of many published reports, the broad range of antigens or toxins proposed to cause these disorders, the lack of definition in terms for current classification schemes combined with an audience that appears to fail to fully understand current disease organization, the concept of a single pathologic cascade appears elusive

Our issue is not with either the authors or the reviewers of this paper who should be commended for their insight in associating marking pens with a LASIK complication. The challenge is that this article was published, presumably after extensive peer review, in one of our profession's most prestigious journals and has yet created no apparent controversy. We believe that this serves to illustrate the level of confusion that appears widespread in our profession on how to appropriately classify and understand the relationship, or lack thereof, between DLK and CTK.

On that note, given the anecdotal nature of many published reports, the broad range of antigens or toxins proposed to cause DLK Stage 4 DLK, CTK, CFN, FNS and CLK, the lack of definition in terms for current classification schemes combined with an audience that appears to fail to fully understand current disease organization, the concept of a single pathologic cascade appears elusive. However, we suggest that the absence of such an understanding will likely serve to perpetuate the current level of confusion and the apparent Tower of Babel nightmare that is a hallmark of this topic. We believe that our model will serve to markedly reduce these issues.

> In the absence of a correct understanding of the pathophysiological cascade, the current level of confusion and the apparent Tower of Babel nightmare that is a hallmark of this topic will likely be perpetuated. Our model will serve to markedly reduce these issues

Histopathologic evidence for current disease paradigms is profoundly lacking

The pathology in both Stage 4 DLK and CTK is currently believed to involve loss of corneal tissue resulting from acute liquefaction necrosis caused by inflammatory cells or tissue toxins[1-6,50]. Unfortunately, these concepts are not supported by published histopathological or confocal microscopy findings[18,52,79,83-92].

> The pathology in both Stage 4 DLK and CTK is currently believed to involve loss of corneal tissue resulting from acute liquefaction necrosis caused by inflammatory cells or tissue toxins. Unfortunately, these concepts are not supported and are, in fact, refuted by published histopathological or confocal microscopy findings. We find this troubling

Buhren et al.[87] described confocal findings in 2 clinical cases of Stage 4 DLK. Pertinent findings included sharp folds in Bowman's membrane, microfolds in the flap stroma, no inflammatory cells and activated keratocytes in the interface. Contrary to conventional thinking, fewer inflammatory cells were found in the interface as DLK progressed. Active inflammation was seen in stages 1 and 2, while stages 3 and 4 revealed the accumulation and decay of inflammatory cells. In addition, the Moilanen et al.[90] confocal study on human subjects found that DLK is not always associated with the presence of inflammatory cells in the interface.

DeRojas Silva et al.[91] described the histopathologic and confocal microscopy findings in rabbit eyes with DLK. Histopathology failed to demonstrate liquefaction necrosis but did show marked edema on both sides of the interface. Confocal microscopy was also obscured by corneal edema but did demonstrate mild infiltration by inflammatory cells as well as abnormal linear structures.

These findings are puzzling considering that the final outcome of Stage 4 DLK is thought to be a result of tissue loss from necrosis and stromal melt. Based on these published studies and reports, strong evidence for liquefaction necrosis of the ECM is lacking. However, ample data does exist to support the occurrence of keratocyte apoptosis in the flap and stroma, tissue edema, a mild inflammatory cell response to keratocyte apoptosis as well as keratocyte and fibroblast activation. Equally perplexing is the recognition that these histopathologic and confocal findings are not consistent with parallel observations of tissue necrosis from infectious keratitis or autoimmune disease[56].

> Current inattention to reconciling actual observed histological findings and known physical properties and physiological responses of the ECM to injury with the clinical disease definition and staging of DLK and CTK is of significant concern

In addition, current disease paradigms for DLK and CTK appear to fail to differentiate between the pathologic effects of keratocyte injury and apoptosis on corneal physiology compared to those affects currently attributed to loss or collapse of the proteoglycan / glycosaminoglycan / collagen matrix. Current inattention to reconciling the actual observed histological findings and known physical properties and physiological responses of the ECM to injury with the clinical disease definition and staging of DLK and CTK is of concern.

Pachymetric evidence for current disease paradigms is equivocal and speculative

Given the pervasive acceptance of stromal necrosis as the etiology for the majority of clinical pathology in these disorders, it may be surprising to recognize that ultrasound pachymetric support for current disease paradigms barely exists. With respect to CTK, CFN, FNS and CLK there are no published cases with collaborating ultrasound pachymetry data in the scientific literature.

> Given the pervasive acceptance of stromal necrosis as the etiology for the majority of clinical pathology in these disorders, it may be surprising to recognize that defensible ultrasound pachymetric support for current disease paradigms for these disorders simply does not exist

Pachymetric evidence for tissue necrosis in Stage 4 DLK is extremely limited. Three published reports in the literature on Stage 4 DLK suggest that corneal thickness is immediately decreased during Stage 4 DLK but becomes thicker during resolution of the disorder[52,53,93].

In these studies, no systematic attempt was made to ensure point-to-point correspondence for serial thickness measurements. Firstly, we suggest that in order to be accepted as legitimate, sequential pachymetric measurements over time used to support or refute a concept of tissue thinning or thickening caused by corneal healing should be mapped to the same location on the cornea. The authors of these studies also reported only single values for ultrasound pachymetry, suggesting that this represented the total change of the entire cornea rather than generating a spatially mapped grid of thickness values across the cornea or even in the central 4 mm. In addition, in no study was there an attempt made to provide data on the variability of sequential pachymetric readings taking during one setting. Also, no study actually measured the cornea by ultrasound pachymetry before and after the development of Stage 4 DLK, but rather focused on tissue recovery during healing.

> **To be accurate, pachymetry** measurements should:
>
> • Ensure point-to-point registration for serial thickness measurements
> • Compensate for the change in speed of sound in edematous cornea
> • Compensate for tissue compression

Of additional concern is the potentially flawed nature of such pachymetric measurements. A seemingly underappreciated fact is that ultrasound pachymetry systems actually only measure the time that transpires between tissue echoes, not the actual distance between tissues[94]. Sound travels through tissues at a rate that is dependent upon the specific sound velocity of that tissue. Knowledge of the average tissue sound velocity for a specific type of tissue can then be used to calculate the thickness of the tissue under consideration using that

known velocity constant. These velocity constants are based upon measurements taken from healthy normal corneas. However, the sound velocity of tissue is dependent upon the hydration state and water content of the tissue. If corneal tissue is highly necrotic or edematous, we believe that the fundamental assumption that its tissue sound velocity is unchanged compared to normal cornea is dubious. As a result, we are skeptical of the validity of such measurements during the acute disease process.

Measurement of ultrasound pachymetry also involves applying sufficient pressure on the corneal surface to obtain a definitive reading. In a tissue substrate that is locally edematous, inflamed and possibly necrotic it is likely that the pressure involved in obtaining the measurement would create tissue compression and distortion. Corneal thickness measurement obtained in this manner are likely invalid and unreliable.

> The pachymetric basis for the premise of stromal loss is poorly substantiated, unproven and highly speculative. In point of fact, no published human histological studies, no published human confocal examinations, no intraoperative observations supported by video or photographic data, no optical coherence tomography data and no published corneal transplant data actually demonstrate loss of corneal volume caused by matrix liquefaction necrosis

If we look beyond pachymetry data for scientific support regarding the concept of tissue necrosis the situation does not improve. Currently, no published human histological studies, no published human confocal examinations, no intraoperative observations supported by video or photographic data, no optical coherence tomography data and no published corneal transplant data actually demonstrate loss of corneal volume caused by matrix liquefaction necrosis.

As a result, the basis for the premise of stromal loss appears poorly substantiated and unproven. Moreover, in the absence of a large sample of ultrasound pachymetry data collected in a disciplined manner that recognizes the limitations of this technology combined with the profound lack of collaborative high frequency ultrasound or optical coherence topographic data, we assert that the notion that there is actual tissue necrosis and ECM loss in the central cornea in Stage 4 DLK, CTK, CFN, FNS and CLK should be viewed as being highly speculative. We also present unequivocal data acquired by OCT and Pentacam imaging technology that completely refutes these concepts.

The OCT and Pentacam imaging studies presented in Figure 9, 10 and 11 of this paper completely refute the concept that there is loss of stroma or that there is clinically relevant tissue necrosis in Stage 4 DLK, CTK and related entities

Conflicting role of interface inflammation in Stage 4 DLK and CTK

A substantial portion of the apparent confusion regarding the pathophysiologic basis for Stage 4 DLK and CTK appears to be dependent upon understanding the role and function of interface inflammatory cells in the pathologic cascade. In fact, a fundamental differentiating feature between Stage 4 DLK, CTK, CFN, FNS and CLK appears to be the ability of the clinician to observe via the slit lamp at some time postoperatively the presence or absence of sufficient numbers of inflammatory cells in the interface subjectively judged to be capable of initiating central tissue necrosis[22]. The precise criterion to be used for this subjective determination

An improved understanding of the precise role of acute inflammatory cells in the interface and their particular function in the induction of pathology in the central cornea would assist in determining the relationship, if any, between these seemingly disparate disorders

apparently remains undefined. Of additional concern is that such a scientifically naive approach to inflammatory disease cascades appears to simply ignore the complex relationship between inflammatory cytokines, inflammatory cells, keratocyte apoptosis and corneal tissue response[7,14-21,42,95-99].

Moreover, according to Sonmez and Maloney[22] there is a very high frequency of association between DLK and CTK. In fact, they suggest that "stage IV lamellar keratitis"...shares no clinical features with diffuse lamellar keratitis except for its frequent coexistence."[22]. Rather than one clinical entity, Sonmez and Maloney[22] appear to suggest that eyes afflicted by CTK almost invariably demonstrate two independent disease processes simultaneously. In our view, this position is nonsensical.

> Ignored by proponents of either disease paradigm is a firm understanding of the role of the inflammatory cell in the disease process. Similarities between the clinical presentation of these disorders suggest that an improved understanding of the precise role of acute inflammatory cells in the interface and their particular function in the induction of pathology in the central cornea would assist in determining the relationship, if any, between these seemingly disparate disorders

Apparently ignored by proponents of either disease paradigm is a firm understanding and recognition the actual role of the inflammatory cell in the overall disease process. Based upon the level of similarities between the clinical presentation of these disorders, it seems reasonable that an improved understanding of the precise role of acute inflammatory cells in the interface and their particular function in the induction of pathology in the central cornea would

assist in determining the relationship, if any, between these seemingly disparate disorders.

We suggest that proponents of the competing paradigms of Stage 4 DLK and CTK are victims of the several common logical fallacies. We agree with Sonmez and Maloney[22] that the idea that inflammatory cells cause Stage 4 DLK primarily suffers from the logical fallacy "Post Hoc, Ergo Propter Hoc". Concluding that DLK causes Stage 4 DLK because it occurred before it temporally currently lacks sufficient evidence to actually warrant such a claim. Similarly, the logical error of "Cum hoc, Ergo Propter Hoc" mistakes correlation for causation, irrespective of their sequential time relationship. It is often easy to believe that because two things occur simultaneously, one must be a cause of the other. However, there is a fundamental difference between correlation and causation.

> Proponents of the competing paradigms of Stage 4 DLK and CTK are victims of the several common logical fallacies. Of most serious concern is that both the Stage 4 DLK and CTK paradigms succumb to the logical fallacy known as "Ignoring a Common Cause"

We also suggest that the idea promoted by Sonmez and Maloney[22] that CTK is caused by photoactivated toxins suffers from causal oversimplification or the "fallacy of the single cause". In this case, Sonmez and Maloney[22] assume that there is one simple cause for a particular disorder when in reality it is likely to be caused by a number of only jointly sufficient causes. Conjoint possibilities are ignored when in fact many different causes or factors may have simultaneously and jointly contributed to the physiologic endpoint.

> If these inflammatory cells are viewed as a symptom of the disease rather than the primary causative factor, their relevance in disease classification is markedly altered

Perhaps of most serious concern is that both the Stage 4 DLK and CTK paradigms appear to succumb to the logical fallacy known as "Ignoring a Common Cause". This fallacy is committed when it is concluded that one thing causes another simply because they are regularly associated. It differs from the previous fallacies in that the causal relationship between toxins and CTK or inflammatory cells and Stage 4 DLK is assumed without consideration of the possibility that a third factor might be the cause of both symptoms. By analogy, a woman is observed standing in a hay field while sneezing and suffering from watery eyes. Proponents of the Stage 4 DLK or CTK paradigms would suggest that the sneezing caused the watery eyes or that the watery eyes caused the sneezing confusing a symptom for causation or that the sneezing is one disease while the watery eyes is another. Both ignore the fact that the woman is standing in a hay field and is suffering from an allergic reaction to airborne pollen. We posit that they have both fallen prey to the logical fallacy of ignoring the common cause.

Specifically, we suggest that if these inflammatory cells are viewed as a symptom of the disease rather than the primary causative factor, their relevance in disease classification is markedly altered. In general, in most diseases, various symptoms may be found in some clinical presentations while absent in others and may wax or wane during the course of the illness. Unfortunately, in the Stage 4 DLK paradigm, such inflammatory cells are speculatively viewed as the immediate cause for the major pathology of the disorder. Conversely, in the CTK paradigm they are viewed as a frequent association of 'excess baggage' that apparently has nothing to do with the final disease pathogenesis. We suggest that both positions are in error because they fail to appreciate the common cause.

> Our model will demonstrate that inflammatory cells in the LASIK interface are a symptom of a single pathologic cascade. They are not the primary cause of the major pathology nor are they a cause of significant tissue necrosis

Our model will demonstrate that inflammatory cells in the LASIK interface are a symptom of a single pathologic cascade. They are not the primary cause of the major pathology nor are they a cause of significant tissue necrosis. However, unlike Sonmez and Maloney, we do not discount the significance or relevance of these cells in the disease process. Rather than distinct entities, we propose that DLK, Stage 4 DLK and CTK are the same clinical entity, as are CFN, FNS and CLK. However, we vigorously reject the pathophysiologic paradigm currently promoted by any of these perspectives. Based on our model, neither the current Stage 4 DLK nor the CTK necrosis paradigm accurately describe the pathophysiologic events that occur in the cornea.

> DLK, Stage 4 DLK and CTK are the same clinical entity, as are CFN, FNS and CLK. However, we vigorously reject the pathophysiologic paradigm currently promoted by any of these perspectives. Neither the current Stage 4 DLK nor the CTK necrosis paradigm accurately describe the pathophysiologic events that occur in the cornea. When faced with multiple competing theories, Occam's principle recommends selecting the theory that introduces the fewest assumptions, postulates the fewest disease entities, relies upon the most balanced logic and is best supported by basic and clinical science data

Rather than accepting a convoluted process where two diseases routinely occur in association with one another or where some unspecified and possibly unknowable aliquot of inflammatory cells determines the fate of the cornea, we suggest the principle of Occam's Razor. When faced with multiple competing theories, Occam's principle recommends selecting the theory that introduces the fewest assumptions, postulates the fewest disease entities, relies upon the most balanced logic and is best supported by basic and clinical science data.

Important clinical observations assist in clarifying the underlying pathologic processes

We believe that our clinical observations on this series of patients demonstrate several under-appreciated facets of the presentation of Stage 4 DLK. Several of these observations were described in an earlier section of this paper. Moreover, these clinical observations provide important clues that assist in clarifying the underlying pathologic process that occurs in these disorders.

Specifically, in each case there was a marked disparity between the degrees of corneal whitening observed using the slit lamp compared to the amount of corneal opacification noted under direct observation in normal room illumination. In our experience with corneal opacification associated with tissue necrosis resulting from infectious keratitis or other causes besides DLK, it is relatively easy to see corneal whitening with even casual external examination in these disorders. By contrast in this series, the affected corneas appeared essentially normal under direct observation. This disconnect in the characteristics of observations made between different examination techniques suggests that the intense whitening is a visual artifact created by scattering of light back toward the slit lamp observer. We find this to be similar to what we observe in Fuch's corneal dystrophy and other disorders that cause corneal edema.

A second important observation was the presence of epithelial bullae and microcystic edema in our most seriously affected and most myopic patients (Case #1 and #3). These findings are typical for a variety of disorders affecting the state of hydration of the cornea.

As a result, we propose that these two clinical observations suggest that the central white opacification observed in Stage 4 DLK, CTK, CFN, FNS and CLK may be related to a dysfunction with respect to the maintenance of normal fluid homeostasis in the stroma rather than tissue necrosis.

> It is critically important to recognize that the clinical phenomenon of late onset DLK strongly suggests that the primary cause of DLK and its related entities, is the release and action of acute inflammatory cytokines in the eye

Association between DLK and epithelial defects is highly problematic

Will first described "late onset DLK" associated with an epithelial defect occurring in a patient several months following uncomplicated LASIK (BR Will MD. Late Onset Interface Inflammation Syndrome. Invited Paper. Ophthalmology 2000...A Higher Calling. American College of Eye Surgeons, QS

XIII. Whistler, British Columbia, Canada. February 14, 1999). Other authors have subsequently also described an association between DLK and epithelial defects occurring weeks or months following LASIK thereby confirming the existence of late onset presentations of this disorder[61,62,70-78].

> Although antigens in the eye may be a factor stimulating inflammation and cause the release of a host of inflammatory chemokines, a specific antigen or toxin deposited in the LASIK interface is a completely unnecessary element to the development of DLK

In the context of current paradigms based upon the notion that an interface antigen drives an acute inflammation process or a toxic substance instilled into the interface results in central tissue necrosis, the occurrence of DLK in association with epithelial defects when no inflammation or toxic necrosis occurred following the initial procedure is highly problematic. Apparently, in these cases the cause of the DLK is the epithelial defect rather than an antigen or toxin. As a result, some authors have proposed a parallel mechanism to antigen deposition in DLK by suggesting a pathological cascade that involves release of inflammatory cytokines by injured epithelial cells, keratocyte apoptosis and an influx of acute inflammatory cells with subsequent tissue necrosis initiated by lytic enzymes and MMP's[42,80,100].

We believe that it is critically important to recognize that the clinical phenomenon of late onset DLK strongly suggests that the primary cause of DLK and its related entities, is the release and action of acute inflammatory cytokines in the eye. In that context, although antigens in the eye may be a factor stimulating inflammation and cause the release of a host of inflammatory chemokines, a specific antigen or toxin deposited in the LASIK interface is a completely unnecessary element to the development of DLK. We suggest that failure to appreciate this relationship has created the unfortunate circumstance wherein the scientific literature on this topic has embraced the logical fallacy of "Ignoring the Common Cause". If a patient presents with a

> We believe that if we recognize that inflammatory cytokines are the primary cause of these disorders, this concept will provide a unifying paradigm that will serve to reduce the confusion currently present in the understanding of the pathophysiologic mechanism of these disorders

complaint of chest pain and shortness of breath, it would be poor medical practice to presume that the chest pain caused the shortness of breath or that the shortness of breath induced the pain in the patient's chest. It is imperative to not confuse symptoms with disease etiology. This latter logically flawed anecdotal thinking accounts for many of the folk lore medical cures rampant in medieval medicine. Today, a trained physician would consider that both symptoms might be caused by coronary artery disease or a pulmonary embolus.

Similarly, we believe that it is reasonable to consider that the inflammatory cells that accumulate in the interface as a symptom of the cascade of pathophysiologic events that occur in response to the effects of inflammatory cytokines and chemokines on corneal tissue and the immune system rather than a cause of DLK, CTK, CFN, FNS or CLK.

We believe that if we recognize that inflammatory cytokines are the primary cause of these disorders, this concept will provide a unifying paradigm that will serve to reduce the confusion currently present in the understanding of the pathophysiologic mechanism of these disorders.

Summary

We believe there to be significant scientific weaknesses in the commonly proposed pathophysiologic mechanisms for DLK, Stage 4 DLK, CTK, CFN, FNS and CLK. Moreover, we have illustrated several logical fallacies associated with these ideas including:

1. Fallacy of the False Premise
2. Post Hoc, Ergo Propter Hoc
3. Cum Hoc, Ergo Propter Hoc
4. Fallacy of the Single Cause
5. Fallacy of Ignoring a Common Cause
6. Fallacy of Reliance on the Highly Implausible Hypothesis
7. Fallacy of Anthropomorphism

Despite such deficiencies, as well as previously articulated weaknesses in data or data interpretation, a review of the literature demonstrates broad unified support for these concepts. Irrespective of this apparent widespread support, we vigorously reject these pathophysiologic models. We suggest that a Chinese proverb may be instructive in understanding what we perceive to be our profession's current predicament.

Pang Cong, an official of the state of Wei was to leave on a trip to the State of Zhao. Before embarking on the trip, Pang Cong asked the King of Wei whether he would hypothetically believe in one civilian's report that a tiger was roaming the markets in the capital city. The King replied "No". Pang Cong asked what the King might think if two people reported the same thing. The King replied "I would begin to wonder if the story was true". Pang Cong then asked, "What if three people all claimed to have seen a tiger?" The King replied "I would believe the story". Pang Cong reminded the King that the notion of a live tiger in a crowded market was absurd, yet when repeated by numerous people, it could seem to be real. As a high-ranking official, Pang Cong had more than three opponents and critics. Naturally, he urged the King to pay no attention to those who would spread rumors about him while he was away. "I understand" the King replied and Pang Cong left for Zhao. While Pang Cong was absent from the capital city, slanderous talk took place. When Pang Cong returned to Wei, the King refused to see him.

Argument ad populum or the notion that "three men make a tiger" is a common fallacious argument that concludes a proposition to be true because many or all people believe it. Unfortunately, history recounts the fact that many people passionately believed that the world was flat, the sun revolved around the earth and that disease was caused by bad ethers.

We posit that the commonly held pathophysiologic model for DLK, Stage 4 DLK, CTK, CFN, FNS and CLK that invokes stromal necrosis as a essential mechanism represents an unfounded and fundamentally flawed premise. Based on our clinical experience and analysis of the associated science, we vigorously suggest that this paradigm involves faulty reasoning and a misinterpretation of reality

The US television comedian, Stephen Colbert coined the term "Truthiness" as a satirical term to describe things that a person claims to know intuitively or "from the gut" without regard to evidence, logic, intellectual examination or facts. Colbert suggested that the word "truth" did not adequately describe the idea that *truth* could be subjective and relativistic. With respect to "truthiness" Colbert declared, "We're not talking about truth, we're talking about something that seems like truth – the truth we want to exist". (S Colbert. The Colbert Report. Pilot Episode. October 17, 2005 and Word of the Year. Merriam-Webster. 2006).

We posit that the commonly held pathophysiologic model for DLK, Stage 4 DLK, CTK, CFN, FNS and CLK that invokes stromal necrosis as a essential mechanism represents an unfounded and fundamentally flawed premise. However, it appears that if this urban legend is mentioned and repeated by many individuals, the premise will be erroneously accepted as the truth.

Moreover, we suggest that the "necrosis paradigm" is a *truth* that we desperately want to believe primarily because it is simple, intuitive, easy to understand and seems "natural". In its unsophisticated splendor, that simplicity is illusory and seductive. Tragically, based on our clinical experience and analysis of the associated science, we vigorously suggest that *this truth* involves faulty reasoning and a misinterpretation of reality.

Unifying Pathophysiologic Model

Overview of the Model

DLK, Stage 4 DLK, CTK, CFN, FNS and CLK are all caused by a loss of control of those factors regulating interstitial fluid pressures (Pif) in the cornea. Inflammation, inflammatory mediators, toxins and cell injury play a role in the etiology of these disorders but do so via their effects on the modulation of Pif.

> DLK, Stage 4 DLK, CTK, CFN, FNS and CLK are all caused by a loss of control of those factors regulating interstitial fluid pressures (Pif) in the cornea. Inflammation, inflammatory mediators, toxins and cell injury play a role in the etiology of these disorders but do so via their effects on the modulation of Pif

Alteration in Pif creates a cascade of physiologic processes that reorder corneal fluid dynamics, alter tissue compliance, redistribute tissue tension across the cornea and initiate reversible biologically mediated mechanical processes. Tissue, ECM or cell necrosis are not significant factors in the etiology or pathology of these disease entities. Moreover, our model suggests that these different diseases exhibit are the product of a single common final pathway or pathophysiologic cascade. As a consequence, we believe that our model will demonstrate that they all represent the same clinical entity.

At the outset, we recognize that this model may appear to be substantially more complicated than the currently accepted but flawed necrosis paradigm. Unfortunately, in the world of physiologic systems, reality is often highly complex. However, we believe that once the basic elements and components of the model are understood, it will be recognized that, in contrast to competing pathophysiologic models, our model does not rely on magical events or unfounded non-physiologic concepts.

Universally recognized properties of corneal tissue include its unique transparency and the relationship between that characteristic and the need to maintain negative interstitial fluid pressures, compact collagen periodicity, tight junctions and constant ion pumping.

> Alteration in Pif creates a cascade of physiologic processes that reorder corneal fluid dynamics, alter tissue compliance, redistribute tissue tension across the cornea and initiate reversible biologically mediated mechanical processes. Tissue, ECM or cell necrosis are not significant factors in the etiology or pathology of these disease entities

Corneal Pif is highly negative with the

center of the cornea being nearly 40 to 50X more negative than dermis. This factor significantly contributes to optical clarity and relatively dehydrated state of corneal tissue. A negative Pif is also essential for the success of LASIK surgery as this is the principle force holding the flap in position following the procedure. Most importantly, the distribution of Pif is not uniform with the central cornea being 30 to 40 X more negative than the limbus

> There is a broad base of scientific evidence that supports the premise that Pif is primarily controlled by specific properties of interstitial cells. Moreover, this body of scientific literature demonstrates that the modulation of Pif is markedly influenced by inflammatory cytokines

and at least 3-5X more negative than peripherally in the cornea (Figure 1). The localization of the most negative Pif at the corneal center provides the visual axis with the most compact collagen / proteoglycan / glycosaminoglycan (GAG) matrix and the highest level of optical clarity. There is a broad base of scientific evidence that supports the premise that Pif is primarily controlled by specific properties of interstitial cells. As a consequence, it is generally well accepted that cell mediated tension forces created by the intracellular actin cytoskeleton transmitted through transmembrane proteins such as ß-integrins to collagen, laminin and fibronectin fibers in the ECM is the principle mechanism involved in the

control of Pif in the body. Moreover, this body of scientific literature demonstrates that the modulation of Pif is markedly influenced by inflammatory cytokines. This includes well known inflammatory mediators including lipopolysaccharides (LPS), prostaglandin derivatives, interleukins (IL), tumor necrosis factor (TNF) and platelet activating factor (PAF). All of these factors have been previously implicated in the pathogenesis of DLK by other authors. As a result, we posit that inflammatory cytokines are a principle mechanism contributing to the causation of DLK, Stage 4 DLK, CTK, CFN, FNS and CLK and that the primary mechanism of this causation is through the modulation of local corneal Pif.

> We posit that inflammatory cytokines are a principle mechanism contributing to the causation of DLK, Stage 4 DLK, CTK, CFN, FNS and CLK and that the primary mechanism of this causation is through the modulation of local corneal Pif

In addition, it has been demonstrated experimentally that directional bulk fluid transport in the rabbit cornea moves fluid from the periphery to the central cornea and is estimated to occur at the rate of 120 nl/hr in the normal state. This property of fluid flow directionality effectively makes the center of the cornea somewhat analogous to the center of a physiologic "black hole".

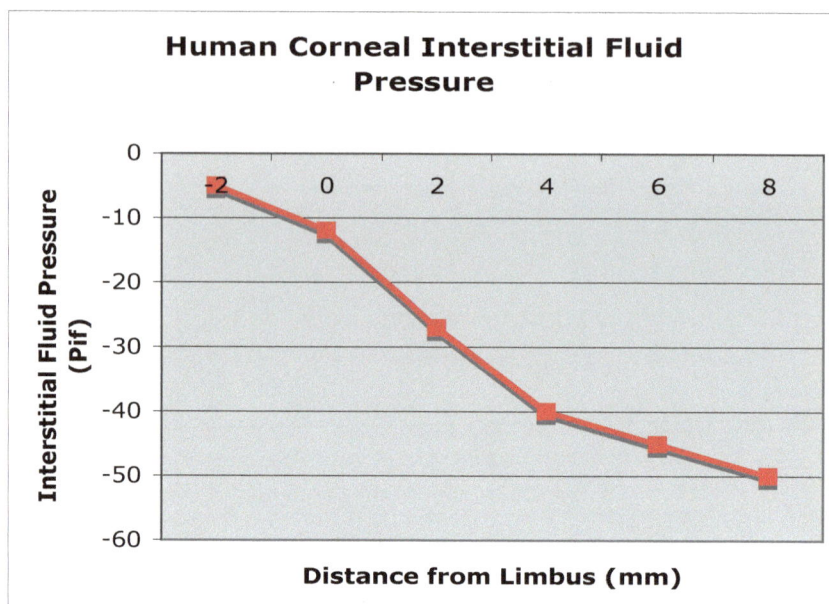

Human Corneal Interstitial Fluid Pressure

Y-axis: Interstitial Fluid Pressure (Pif)
X-axis: Distance from Limbus (mm)

Figure 1 - Interstitial fluid pressure in the Human cornea

Interstitial fluid pressure (Pif) in the cornea is not distributed in a homogenous fashion. The central cornea is 30 to 40 X more negative than the limbus, is at least 3-5 X more negative than peripherally in the cornea and is nearly 40 to 50 X more negative than dermis. The localization of the most negative Pif at the corneal center likely provides the visual axis with the most compact ECM and the highest level of optical clarity. This Pif gradient creates directionality in bulk fluid flow from periphery to corneal center. We suggest that this Pif gradient and the mechanisms that control Pif and influence fluid flux in the cornea are the primary drivers causing KME. A mechanism for creation of either the Pif gradient or the directionality of bulk fluid flow is not understood. We believe that the Pif gradient may be attributed to limbal fluid influx, peripheral collagen interweaving that limits tissue expansion and serves to slow fluid movement from the limbus, gradients in endothelial and keratocyte cell populations from periphery to center and localized endothelial pump effects. We also define a mechanism wherein there is a decrease in the local effect of the endothelial pump as tissue becomes located further and further away from the endothelial layer. Conversely, tissue located closest to the endothelial cell layer would exhibit an increased local pump effect. Since the cornea is thinnest near the center, spatially mapped variance in pump efficiency caused by change in corneal thickness could explain the Pif gradient and fluid movement directionality. We suggest that the rate at which pump efficiency decreases with distance may be linear or non-linear. (Redrawn from Wiig[166])

Likely as a consequence of the marked Pif gradient, corneal fluid dynamics are unidirectional and, under normal physiologic conditions, the arrow of that directional marker consistently points towards the corneal center. Our model suggests that fluid movement from limbus to corneal center occurs in a manner that respects the anisotropic properties of corneal structure and anatomy wherein fluid moves towards the center by primarily following the path of least resistance along hydrogel 'columns' between collagen lamellae. The combination of the Pif gradient and this anisotropic flow pattern creates "fan shaped" flow trajectories wherein the flow of fluid from limbus to corneal center follows an "arc-shaped" pattern similar in shape to an inverse parabola (Figure 2).

We assert that control of Pif in the cornea is disrupted during laser refractive surgery by essentially two different mechanisms or cascades converging on a final common pathway (Figure 3). We define an Inflammatory Arm (or 'DLK arm') and a Non-inflammatory Arm (or 'CTK arm'). Irrespective of this distinction, our model recognizes that in a post refractive surgery cornea, numerous inflammatory and non-inflammatory mechanisms co-exist and considerable overlap occurs in the conjoint mechanisms directing the pathogenesis of a single final disease entity.

The Inflammatory Arm (or 'DLK arm') of the pathophysiologic cascade includes the effects of inflammatory cytokines and mediators including, but not limited to, various forms of IL, TNF, PAF, LPS and prostaglandin derivatives that bind to

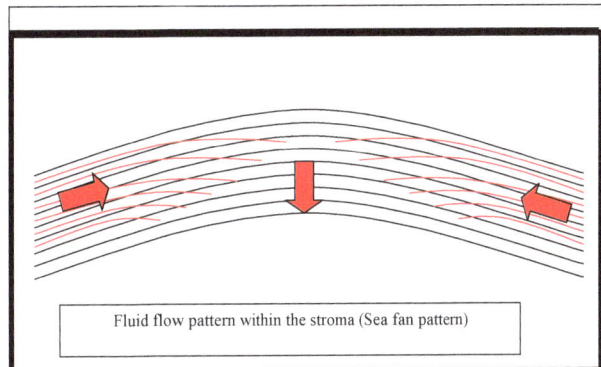

Fluid flow pattern within the stroma (Sea fan pattern)

Figure 2 - Anisotropic fluid flow

We posit that fluid movement from periphery to center in the cornea follows an anisotropic "fan shaped" pattern. This fan shaped or inverse parabolic fluid trajectory is created by the "onion-like" anisotropic properties of corneal anatomy and structure. Fluid physically tracks more easily in a direction parallel to layers of collagen mesh than perpendicular to such tightly packed arrays. By analogy, it is much easier to drive a nail "with" the grain of a piece of wood than "across" the grain. As fluid moves towards the corneal center, traveling from periphery to center, it is progressively pulled towards the posterior of the cornea and is actively pumped out of the center of the cornea due to the increasing negativity of the Pif gradient in the corneal center and the local influence of the endothelial pump mechanism as fluid and tissue become progressively more proximal to this pumping process. This serves to overcome fluid layering created by anisotropic tissue characteristics. The combination of Pif gradient and anisotropic fluid movement creates a "conveyor belt" effect wherein fluid is moved primarily to the corneal center where it is pumped into the anterior chamber.

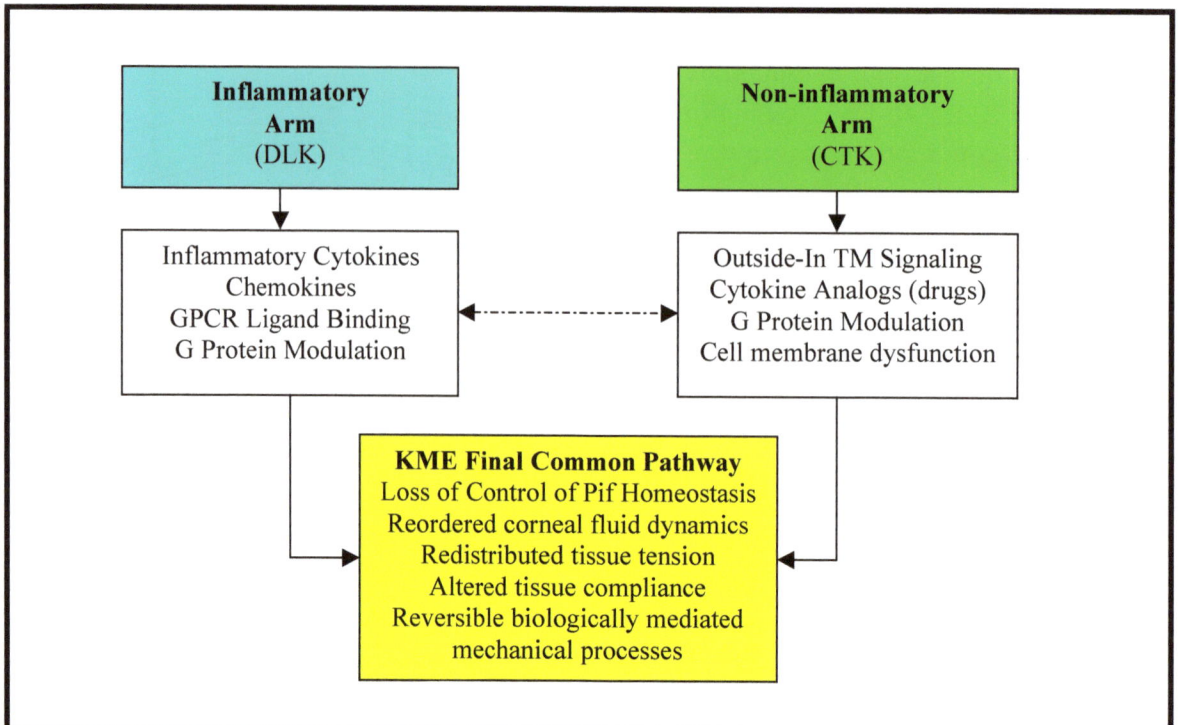

Inflammatory Arm (DLK)	Non-inflammatory Arm (CTK)
Inflammatory Cytokines Chemokines GPCR Ligand Binding G Protein Modulation	Outside-In TM Signaling Cytokine Analogs (drugs) G Protein Modulation Cell membrane dysfunction

KME Final Common Pathway
Loss of Control of Pif Homeostasis
Reordered corneal fluid dynamics
Redistributed tissue tension
Altered tissue compliance
Reversible biologically mediated
mechanical processes

Figure 3 - Pathophysiologic Model for KME

The pathophysiologic model for KME postulates that DLK, Stage 4 DLK, CTK, CFN, FNS and CLK are all caused by the loss of control of Pif. Control of Pif in the cornea is disrupted during laser refractive surgery by two different mechanisms or cascades converging on a final common KME pathway. The Inflammatory Arm or 'DLK arm' of the pathophysiologic cascade includes the effects of inflammatory cytokines and modulation of G-protein directed cell functions. The Non-inflammatory Arm or 'CTK arm' of the pathophysiologic cascade includes factors that directly affect keratocyte function and viability. Rather than being independent, these mechanisms frequently conjointly contribute to the development of this disorder. Alteration in Pif creates a cascade of physiologic processes that reorder corneal fluid dynamics, redistribute tissue tension across the cornea, alter tissue compliance and initiate reversible biologically mediated mechanical processes. Reversible biomechanical forces cause marked shifts in refraction, transient opacification of central flap stroma and tissue macrofolds.

receptors on keratocyte membranes and create a physiologic effect through the modulation of G-protein directed cell functions (Figure 4).

The Non-inflammatory Arm (or 'CTK arm') of the pathophysiologic cascade includes factors that directly affect keratocyte function and viability (Figure 5). These factors include the modulation of keratocyte function resulting from mechanical ECM-cell "outside-in" β-integrin mediated transmembrane signaling processes, G-protein modulated effects and the disruption of cell membrane function and architecture. In clinical practice, these latter factors include both mechanical and thermal laser-tissue interactions, the effects of pharmacologic agents that bind to G-protein receptors on keratocyte cell membranes and effects of toxic agents such as organic solvents, topical anesthetics and preservatives such as benzalkonium chloride (BAK) that interfere with cell membrane homeostasis and the conformation of transmembrane proteins.

As a consequence of change in function, behavior and viability of keratocytes in the flap and stroma, interstitial swelling pressures are immediately affected through cell mediated and osmotic effects. Specifically, Pif becomes more negative causing the cornea to imbibe fluid. Fluid uptake occurs preferentially in local stromal tissue that is exposed to the effects of inflammatory cytokines. Areas most affected are nearest to the flap sidewall where injured epithelium releases prostaglandins and other cytokines and also represents the path of least resistance

for cytokines and fluid influx into the stroma. In addition, the center of the cornea where Pif is most negative draws fluid from the tear film. As Pif in the corneal center is driven more negative this amplifies bulk fluid movement towards the central cornea and the anterior stromal tissue becomes increasingly edematous.

Acute inflammatory cells may present in the tear film as a result of immune response in the body from iatrogenic injury related to the LASIK procedure, atopic disease or may be recruited by inflammatory cytokines released by excessive epithelial injury from an epithelial defect. If present in the tear film, those inflammatory cells are effectively pulled into the interface by bulk fluid flow. Movement of these inflammatory cells, carried by bulk fluid flow towards the corneal center, occurs along the LASIK interface that acts as a path of least resistance. This creates the pathognomonic concentric "shifting sands" appearance as inflammatory cells are 'sorted' into 'windrows' by the action of waves of bulk fluid pulled to the center of the cornea by the Pif gradient.

These inflammatory cells represent a symptom of the excessive activity of inflammatory cytokines. These cells are not the cause of DLK, Stage 4 DLK, CTK, CNS, FNS or CLK. They also do not release clinically significant amounts of metalloproteinases or collagenases, they do not "clump" together, they do not cause clinically significant tissue necrosis and they are not responsible for whitening of the cornea. However, these inflammatory cells may release or induce the synthesis of additional inflammatory

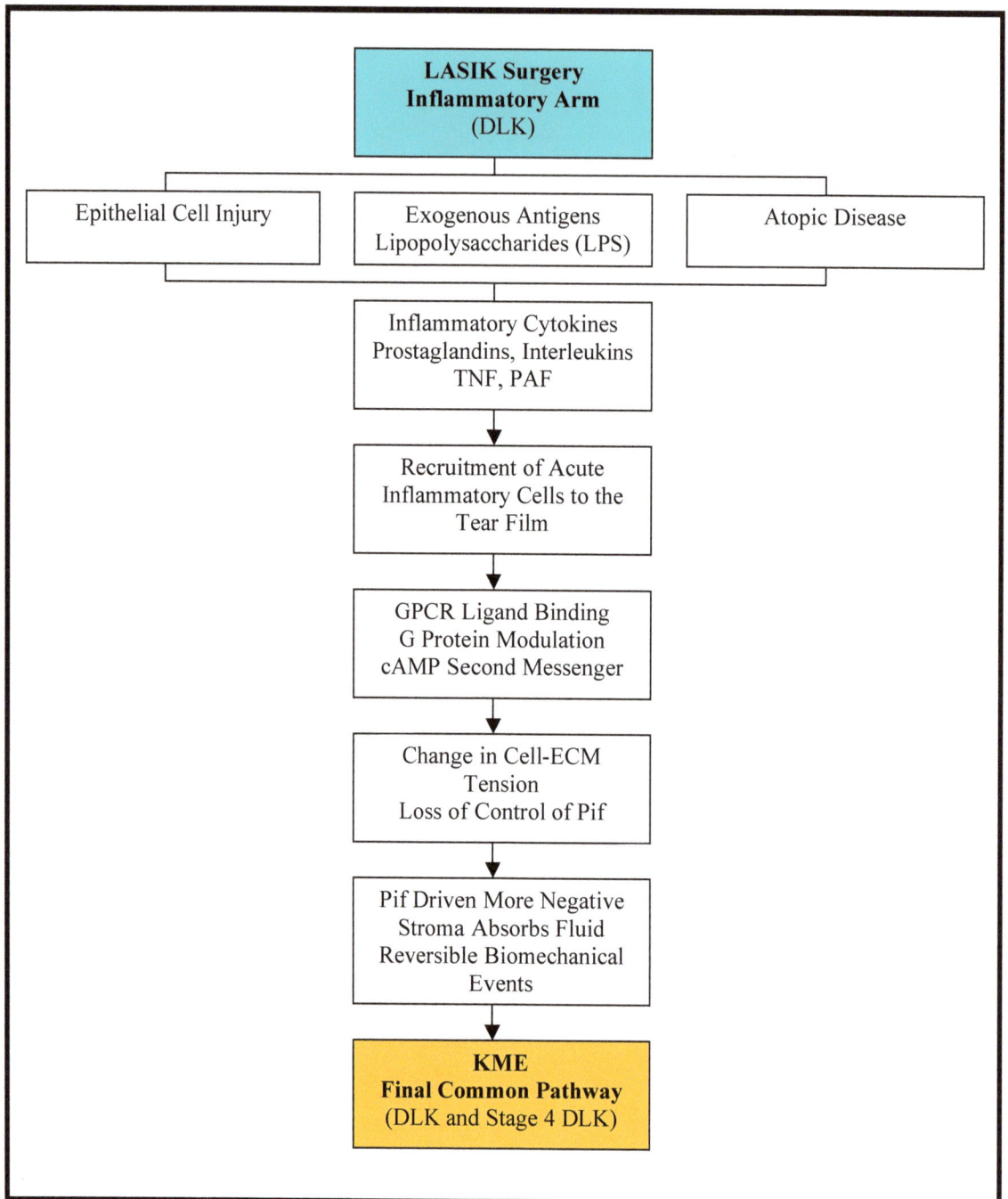

Figure 4 - Inflammatory Arm of KME Pathophysiologic Cascade

The Inflammatory Arm of the pathophysiologic cascade for KME recognizes that LASIK surgery creates inflammation by injuring tissue, introducing exogenous antigens (such as LPS) and accentuating pre-existing inflammatory conditions. These factors induce the synthesis of inflammatory cytokines including prostaglandin derivatives, IL, TNF and PAF that bind to receptors on keratocyte membranes and create a physiologic effect through the modulation of G-protein and cAMP directed cell functions. These inflammatory mediators also recruit acute inflammatory cells to the tear film. The interface acts as a path of least resistance and the "shifting sands" appearance results as inflammatory cells are pulled to the center of the cornea by the Pif gradient. Inflammatory cells are a symptom rather than a cause of the disorder. Inflammatory cytokines induce changes in cell-ECM tension that disrupt local control of Pif. Pif is driven more negative causing the cornea to imbibe fluid. Fluid uptake occurs preferentially in local stromal tissue that is exposed to the effects of inflammatory cytokines. Marked local changes in Pif and tissue tension cause significant alteration in local compliance characteristics of the ECM. The transient central whitening and folding of the flap as well as the hyperopic shift are due to factors affecting the compliance of the corneal extracellular compartment and are the product of Pif induced reversible biomechanical forces. None of these clinical findings are caused by tissue necrosis.

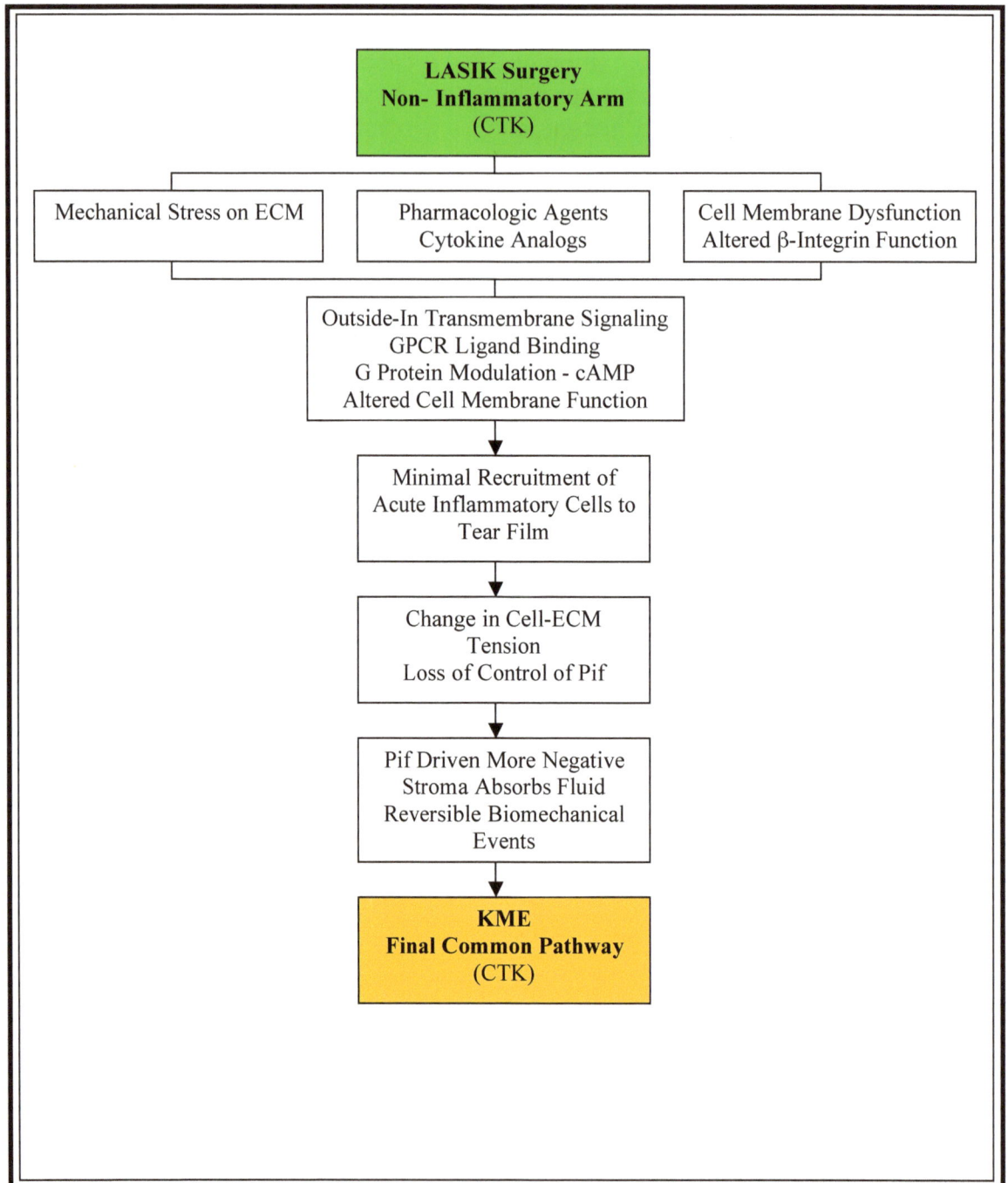

LASIK Surgery
Non- Inflammatory Arm
(CTK)

| Mechanical Stress on ECM | Pharmacologic Agents Cytokine Analogs | Cell Membrane Dysfunction Altered β-Integrin Function |

Outside-In Transmembrane Signaling
GPCR Ligand Binding
G Protein Modulation - cAMP
Altered Cell Membrane Function

Minimal Recruitment of
Acute Inflammatory Cells to
Tear Film

Change in Cell-ECM
Tension
Loss of Control of Pif

Pif Driven More Negative
Stroma Absorbs Fluid
Reversible Biomechanical
Events

KME
Final Common Pathway
(CTK)

Figure 5 - Non-inflammatory Arm of KME Pathophysiologic Cascade

The Non-inflammatory Arm of the pathophysiologic cascade for KME includes those factors in LASIK surgery that directly affect keratocyte function and viability and result in disruption of control of Pif. These factors include the modulation of keratocyte function resulting from mechanical ECM-cell "Outside-In" β-integrin mediated transmembrane signaling processes, G-protein modulated effects and the disruption of cell membrane function and architecture. In clinical practice, these latter factors include both mechanical and thermal laser-tissue interactions, the effects of pharmacologic agents that bind to G-protein coupled receptors on keratocyte cell membranes and effects of toxic agents such as organic solvents, topical anesthetics and preservatives such as benzalkonium chloride that interfere with cell membrane homeostasis. Since these factors do not include the induction of inflammatory cytokines, in this arm recruitment of acute inflammatory cells to the tear film and interface is minimal. The resulting alteration in Pif creates a cascade of physiologic processes that reorder corneal fluid dynamics causing the cornea to imbibe fluid. Marked local changes in Pif and the redistribution of tissue tension across the cornea significantly alter local tissue compliance and initiate reversible biologically mediated mechanical processes.   These reversible biomechanical forces cause marked shifts in refraction, transient opacification of central flap stroma and tissue macrofolds. None of these clinical findings are caused by tissue necrosis.

cytokines that can, in theory, further worsen the change in Pif.

As the extracellular matrix compartment becomes increasingly saturated the separation between collagen fibrils in the LASIK flap markedly increases and corneal clarity begins to decrease, particularly in the LASIK flap where tissue tension is markedly reduced. In Stage 4 DLK, CTK, CFN, FNS and CLK, edema causes intense central focal whitening of the LASIK flap, which is evident under slit lamp examination due to its profound light scattering effects. Severe flap macrostriae occur in the center of the cornea as the result of marked expansion of stromal tissue in the flap due to edema. Histologically, keratocyte cell apoptosis and cell necrosis may occur, however there is minimal, if any, actual loss of the proteoglycan, glycosaminoglycan, collagen matrix in either the flap or residual stromal bed due to liquefaction necrosis.

The hyperopic shift, as well as the central whitening and folding of the flap that occurs in Stage 4 DLK, CTK, CFN, FNS and CLK, are due to factors affecting the compliance of the corneal extracellular compartment. None of these clinical findings are caused by tissue necrosis. Moreover, Pif and tissue tension represent the primary variables that control compliance characteristics of the ECM.

Generally, in an undisturbed tissue with a negative Pif, compliance (ΔVol / ΔPif) follows a more or less linear relationship between change in tissue volume for a particular change in Pif. However, as Pif

becomes positive, the linearity of this relationship can be lost. Depending upon those forces that restrain further tissue expansion, as Pif becomes positive the tissue may take up significantly more fluid and exhibit marked swelling with no further change in Pif. Typically, when the latter occurs the tissue becomes visibly edematous.

In addition to Pif, our model suggests that compliance of the ECM is also markedly affected by tissue tension. As tension increases, the slope of the relationship between ΔVol and ΔPif becomes steeper and compliance of the compartment decreases. If tension is very high the tissue effectively demonstrates little if any compliance and the size of the extracellular compartment may decrease independently from Pif. Conversely, as tissue tension decreases, compliance of the compartment increases such that the amount of change in tissue volume for any given change in Pif becomes larger (Figure 6).

We posit that in Stage 4 DLK, CTK, CFN, FNS and CLK, local changes in Pif and tissue tension induce a marked hyperopic shift as well as create the central whitening and folding of the flap. These clinical events occur due to direct effects of Pif and tissue tension on the local compliance of the corneal extracellular compartment.

The compliance characteristics of the LASIK flap and residual stromal bed in the post-LASIK cornea are markedly different from the preoperative condition. That is because post-LASIK, tension and

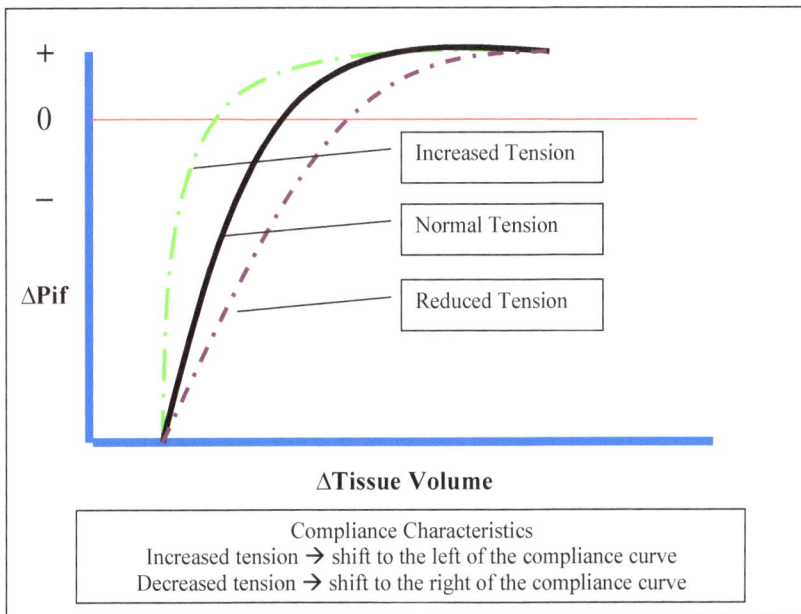

Figure 6 - Tissue Compliance Curve - (ΔVol / ΔPif)

Tissue compliance (ΔVol / ΔPif) follows a linear relationship between change in tissue volume for a particular change in Pif when Pif is negative. However, as Pif becomes positive, the linearity of this relationship can be lost. Depending upon those forces that restrain further tissue expansion, as Pif becomes positive the tissue may take up significantly more fluid and exhibit marked swelling with no further change in Pif. Typically, when the latter occurs the tissue becomes visibly edematous. In addition to Pif, our model suggests that compliance of the ECM is also markedly affected by tissue tension. We posit that as tension increases, the slope of the relationship between ΔVol and ΔPif becomes steeper and compliance of the compartment decreases. If tension is very high the tissue effectively demonstrates little if any compliance and the size of the extracellular compartment may decrease independently from Pif. Conversely, as tissue tension decreases, compliance of the compartment increases such that the amount of change in tissue volume for any given change in Pif becomes larger. (Adapted from Wiig[55]) mechanical forces on the LASIK flap and residual stromal tissue are significantly altered. This has the specific effect of altering the relationship between tension and Pif that normally occurs in a predictable state of homeostatic balance in the pre-LASIK cornea. Changes in tension and Pif also do not occur in a homogenous manner in the post-LASIK cornea. Rather, in this unifying model we identify four (4) zones that undergo distinct changes in compliance (Figure 7 through 12);

1. Zone 1 - The LASIK flap
2. Zone 2 - The anterior cornea that is peripheral to the LASIK flap
3. Zone 3 - The cornea immediately posterior to Zones 1 and 2. We further divide Zone 3 into two (2) sub-zones 3a and 3b based on their unique compliance characteristics
4. Zone 4 – The cornea posterior to Zone 1, 2 and 3

Zone 1: Compared to the preoperative condition, in Zone 1 mechanical tension in the flap is markedly reduced because the vast majority of collagen fibrils connecting the flap to the residual stroma have been severed. In the absence of tension created by collagen fibrils, tissue compliance and $\Delta Vol / \Delta Pif$ are increased. A highly negative Pif causes the corneal flap tissue to imbibe fluid following the new $\Delta Vol / \Delta Pif$ relationship defined by a tissue system operating under reduced tension. As Pif becomes increasingly positive, tissue volume centrally begins to expand at a very rapid rate and may create gross flap edema. This, in combination of a compliance mismatch between Zone 1 and Zone 3b, creates focal whitening of

the cornea and full thickness macrofolds. However, once this central tissue attains saturation, the gradient drawing fluid to the center is significantly reduced. The resulting reduction of the Pif gradient combined with the gradual re-establishment of the tight junctions in the epithelial layer inhibits additional fluid and cell movement towards the corneal center.

Zone 2: In the peripheral anterior zone of the cornea (Zone 2) tension on collagen fibrils and Pif are also decreased. Similar to Zone 1, the decreased Pif and increased compliance of Zone 2 allows this tissue to swell and creates a thickening of the local corneal tissue. This tissue swelling also has no significant impact on tissue strain or tension because the normal span these fibrils has been severed by the flap and excimer laser processes. However, from a refractive perspective, this tissue swelling creates a transient peripheral steepening. This steepening contributes to a hyperopic shift and the induction of positive spherical aberration. However, as Pif becomes more normal, this peripheral swelling and steepening demonstrates reversibility.

Zone 3: The pathologic processes in Zone 3 are all fundamentally driven by changes in Pif. However, the dynamic between mechanical tension and tissue swelling due to changes in Pif is more complex. Alterations in Pif, fluid influx, tension and tissue compliance in Zone 3 cause two related clinical effects;

1. The reversible hyperopic shift. The contribution of Zone 3 is significantly greater than that of Zone 2.

2. A compartment syndrome that traps
   fluid in the central flap (Zone 1)

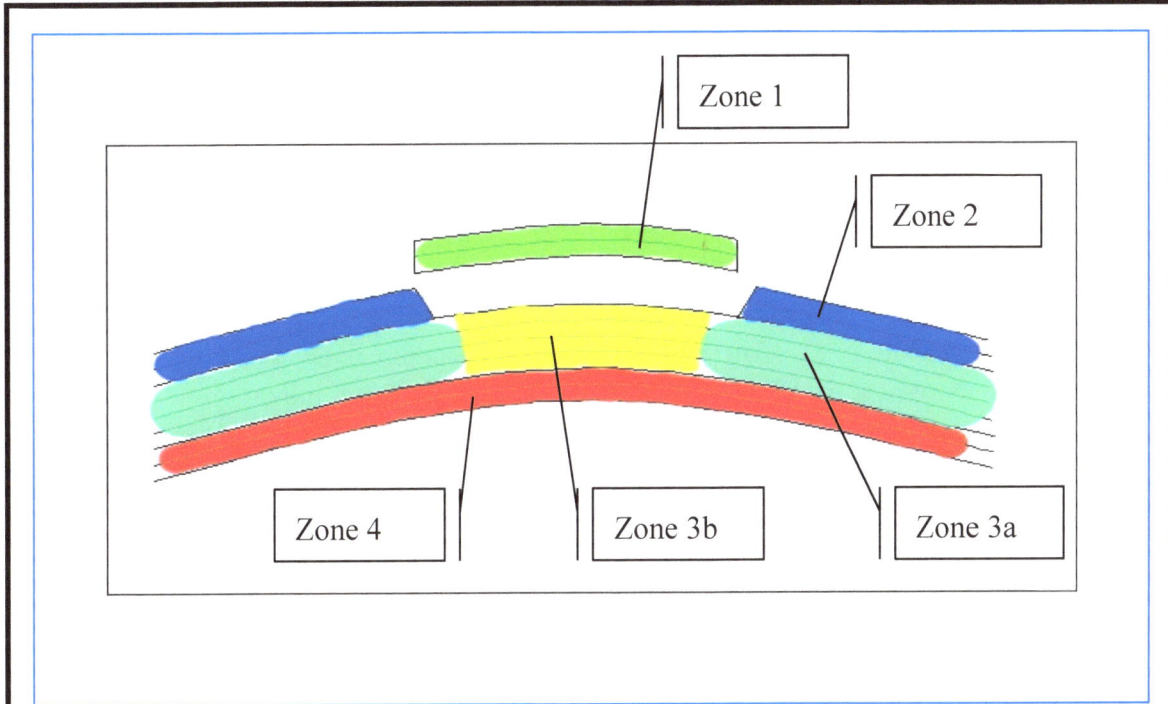

Figure 7 - Compliance Characteristics of the LASIK flap and residual stroma

The compliance characteristics of the LASIK flap and residual stromal bed in the post-LASIK cornea are markedly different from the preoperative condition. That is because post-LASIK, tension and mechanical forces on the LASIK flap and residual stromal tissue are significantly altered. This has the specific effect of altering the relationship between tension and Pif that normally occurs in a predictable state of homeostatic balance in the pre-LASIK cornea. Changes in tension and Pif also do not occur in a homogenous manner in the post-LASIK cornea. Rather, in this unifying model we identify four (4) zones that undergo distinct changes in compliance;

1. Zone 1 - The LASIK flap
2. Zone 2 - The anterior cornea that is peripheral to the LASIK flap
3. Zone 3 - The cornea immediately posterior to Zones 1 and 2. We further divide Zone 3 into two (2) sub-zones 3a and 3b based on their unique compliance characteristics
4. Zone 4 – The cornea posterior to Zone 1, 2 and 3

These two distinct clinical effects are caused by the post-LASIK relationship between tension, Pif and compartment compliance. As a result of these factors, based on tissue response characteristics, two additional distinct sub-zones (Zone 3a and 3b) are created in Zone 3.

Tissue tension dynamics are markedly different in Zone 3 compared to Zone 1 and 2 and the effect of tension on tissue compliance is of major clinical significance. Firstly, injured epithelium surrounding the flap sidewall typically represents the largest source of inflammatory cytokines in the cornea and local disruption in the normal barrier function becomes the primary location for diffusion of these inflammatory cytokines into the local stroma. Secondly, this area of the flap sidewall is also the path of least resistance for fluid and inflammatory cells from the tear film. Thirdly, fluid in the tear film that is drawn into the interface is rapidly absorbed by the corneal stroma near the flap sidewall due to its highly negative Pif caused by these local cytokines. In addition, the increase in negativity in Pif in the flap sidewall region begins to draw significant amounts of fluid from the limbus. As fluid is selectively absorbed by stroma near this circular entry portal, the ECM begins to expand and the tissue swells and thickens. Fluid flowing into the cornea from the limbus and the entrance portal created by the flap sidewall in response to the local decrease in Pif in the area combine to create a preponderance of swelling and thickening in Zone 3a compared to Zone 3b.

As the stroma in Zone 3a becomes increasingly edematous the cornea in that region begins to expand and thicken. Corneal edema and expansion of the ECM in Zone 3a causes the collagen fibrils present there to move further apart. However, the eye is a closed hydraulic system and it is effectively incompressible. As a consequence, as the collagen fibrils move further apart, they expand proceeding outwards away from the center of the eye in a centrifugal direction. During this tissue expansion process, the origins of these collagen fibrils remain fixed in position at the limbus and cannot move. Since collagen is not elastic, this peripheral outward tissue expansion with fixed anchor points at the limbus begins to place increasing tension on those collagen fibrils that remain intact in the central cornea. In order for tissue expansion to occur in Zone 3, the cord length of the collagen fibrils would need to increase, particularly in the fibrils in the outer aspect of Zone 3. An increase in cord length is not possible. As a consequence, the curve created by these collagen fibrils becomes distorted such that they are steepened and flattened in a localized fashion. As the tissue in Zone 3a swells it places progressively more tension on the collagen fibrils in Zone 3b. Without an ability to increase their cord length, this increased tension causes a flattening effect in Zone 3b. At the same time, this swelling creates a localized steepening in Zone 3a.

A simple analogy to assist in understanding such a phenomenon might be to envision the distortion of a dome shaped camping tent if outward pressure is applied. Similar to the cornea, adding to the cord length of the nylon fabric is not possible and the tent base is fixed. Therefore, as the mid peripheral area of the dome tent moves outwards, the center and apex of the dome tent must flatten in a compensatory fashion. This effect results in local mid-peripheral steepening of the dome tent walls, central flattening of the dome of the tent and flattening of the most peripheral tent walls near the tent base.

A simple analogy to assist in understanding such a phenomenon might be to envision the distortion of a dome shaped camping tent if outward pressure is applied. Similar to the cornea, adding to the cord length of the nylon fabric is not possible and the tent base is fixed. Therefore, as the mid peripheral area of the dome tent moves outwards, the center and apex of the dome tent must flatten in a compensatory fashion. This effect results in local mid-peripheral steepening of the dome tent walls, central flattening of the dome of the tent and flattening of the most peripheral tent walls near the tent base

The hyperopic shift in cases of Stage 4 DLK, CTK, CFN, FNS and CLK are caused by this biologically mediated mid-peripheral corneal steepening and central corneal flattening. The swelling of Zone 3a is responsible for inducing local peripheral steepening. Conversely, the flattening and thinning of Zone 3b is also caused by localized thickening of Zone 3a. This effect occurs to some degree in most cases of LASIK. In virtually all cases, the peripheral steepening and central flattening effect is reversed when Pif and fluid dynamics become normalized. It is for this reason that the hyperopic shift in these disorders is reversible.

Increasing tissue tension has the effect of increasing the slope of the compliance curve and decreasing tissue compliance. As a result of stretching and flattening of the central cornea (Zone 3b) the compliance of the Zone 3b compartment is decreased. With increasing tension and decreasing tissue compliance, fluid is actually excluded from the tissue following this new ΔVol / ΔPif compliance curve. Due to this change in tissue compliance, the residual corneal bed in the center of Zone 3b becomes transiently thinner. This thinning of the tissue due to tissue compression further contributes to the marked central flattening and exacerbates the hyperopic shift.

As tissue compliance decreases in Zone 3b, the ability for fluid to move from the flap (Zone 1) into Zone 3b decreases. If severe, the change in compliance creates a compartment block or syndrome wherein fluid moving into Zone 1 becomes trapped and is unable to move into Zone 3b. This effect is further compounded by the high affinity for water by the GAG's in Zone 1 combined with the increased tissue compliance in Zone 1.

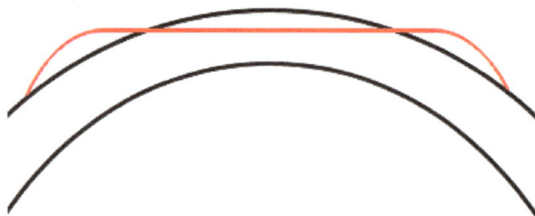

It is this process that causes the central flap to become extremely edematous causing focal central flap whitening, a "cone shaped" white area, flap macrofolds and tissue opacification. The fan shaped inverse parabolic fluid movement pattern created by the anisotropic structure of the cornea determines that this whitening and flap folding occur directly in the corneal center. However, as inflammation and other factors subside in the postoperative period, this compartment block gradually resolves and the central flap whitening and macrofolds slowly fade. There is no clinically relevant necrosis or stromal remodeling. If the patient exhibits loss of BSCVA it is a result of unresolved macrofolds in the flap. Residual microfolds that adversely affect BSCVA must be corrected surgically or visual recovery will be limited or markedly delayed.

Zone 4: Physiologic changes in Zone 4 do not contribute significantly to the clinical manifestation of DLK, Stage 4 DLK, CTK, CFN, FNS or CLK because the cornea is not significantly impacted by events of the surgery. Tissue tension in Zone 4 is not much different than in the pre-surgical state and the effect of inflammatory cytokines is less than in more anterior areas of the cornea. As a result of nominal changes in tissue tension

and Pif, the slope of the compliance curve remains unchanged. The proximity of the tissue to the endothelial pump also remains unchanged from the pre-operative state. This latter observation, combined with the low water binding characteristic of keratan sulphate located most posteriorly in the cornea, allows this area of the stroma to give up its water more readily than in Zone 1, 2 or 3.

In summary, transient mid-peripheral corneal steepening and central corneal flattening cause the hyperopic shift observed in cases of Stage 4 DLK, CTK, CFN, FNS and CLK. They also contribute significantly to a compartment block that traps fluid in the center of the flap. Rather than occurring almost entirely on the basis of mechanical change in collagen tension due to reduction in numbers of intact collagen fibrils, our model suggests that these changes in mechanical forces substantively result from biological changes determined by control of loss of control of local Pif. Since changes induced in control of Pif from refractive surgery can be reversed with tissue healing, the hyperopic shift and central corneal edema so induced also exhibit features of reversibility. Moreover, the reversibility of the effect is clearly documented in the Pentacam™ tomography (Oculus, Bellevue, WA, USA) and Optical Coherence Tomography (OCT) (Zeiss Humphrey®, Dublin, CA, USA) studies included in this paper (Figures 9 through 12).

As the keratocyte population physiology recovers over time, control over Pif is re-established and the tendency of the tissue to imbibe fluid is markedly diminished.

As Pif dynamics become increasingly more normal, the compliance of the three zones in the cornea begin to drift towards a more normal and homeostatic condition. The result of this normalization in control of Pif is that the hyperopic shift and associated induced astigmatism gradually resolves over time. Swelling pressures that create collagen tension centrally and peripherally become more normal. Depending upon the physiologic response of the specific cornea, the hyperopic shift and accompanying astigmatism may resolve completely or may leave the patient with a residual refractive error.

An elevated intraocular pressure has the adverse effect of reducing the speed at which this recovery may occur. If intraocular pressure is elevated from a steroid response, compressive forces transmitted to the residual stromal bed will markedly exacerbate the lack of compliance of the posterior central stroma. As a consequence, fluid located anteriorly in the central stroma cannot physically move in an efficient manner towards the endothelial layer, effectively creating a "compartment" syndrome. In that setting, fluid accumulates in the epithelium in the form of bullae or, if severe, in the interface. The variance in tissue compliance induced by an elevated IOP is the primary mechanism causing interface fluid syndrome (IFS).

Residual flap macrostriae may also largely resolve spontaneously as the edema goes away. However, in some cases residual macrostriae in the visual axis can reduce BSCVA. If present and visually significant, our model and experience recommend a flap lift and stretch procedure to treat such microstriae. Any residual refractive error can be subsequently treated with additional photoablation, however such treatments should be performed sequentially rather than simultaneously.

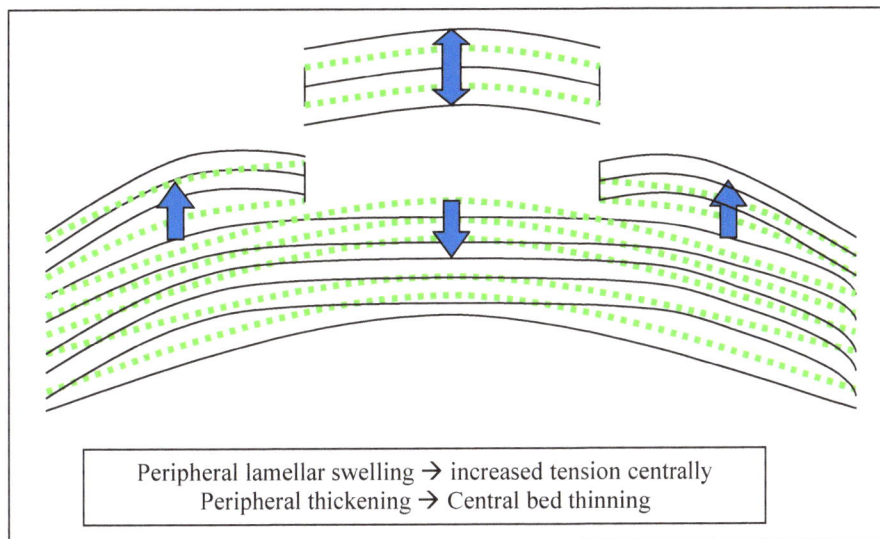

Peripheral lamellar swelling → increased tension centrally
Peripheral thickening → Central bed thinning

Figure 8 - Change in corneal tension due to local edema

Tension in the flap (Zone 1) is markedly reduced causing an increase in tissue compliance. A highly negative local Pif causes the corneal flap tissue to imbibe fluid. As Pif becomes increasingly positive, tissue volume centrally begins to expand at a very rapid rate creating flap edema. In the peripheral anterior zone of the cornea (Zone 2) tension on collagen fibrils and Pif are also decreased. A decreased Pif and increased compliance of Zone 2 allows this tissue to swell creating a thickening of the local corneal tissue and a marked transient peripheral steepening that contributes to a hyperopic shift and the induction of positive spherical aberration. In Zone 3 alterations in Pif, fluid influx, tension and tissue compliance cause a reversible hyperopic shift and a compartment syndrome that traps fluid in the central flap (Zone 1). Inflammatory cytokines infusing into the stroma through the flap sidewall drive local Pif negative and begin to draw significant amounts of fluid from the limbus and tear film. The local ECM begins to expand and the tissue swells and thickens. As the stroma in Zone 3a becomes edematous, tissue expansion and thickening places progressively more tension on the collagen fibrils in Zone 3b. Due to their fixed cord length, this increased tension causes a flattening effect in Zone 3b combined with a localized steepening in Zone 3a. This biologically mediated central corneal flattening, mid-peripheral steepening and peripheral flattening causes the hyperopic shift. This effect is reversed when Pif and fluid dynamics become normalized and it is for this reason that the hyperopic shift in these disorders is frequently reversible. Compression and flattening of the central cornea (Zone 3b) also causes decreased tissue compliance. Increasing tension and decreasing tissue compliance excludes fluid and the residual corneal bed in the center of Zone 3b becomes transiently thinner further contributing to the marked central flattening and exacerbating the hyperopic shift. With decreasing compliance, the ability for fluid to move from the flap (Zone 1) into Zone 3b also decreases. If severe, the change in compliance creates a compartment block or syndrome wherein fluid moving into Zone 1 becomes trapped and is unable to move into Zone 3b. It is this compartment block that causes the central flap to become extremely edematous exhibiting focal central flap whitening, flap macrofolds and tissue opacification. As inflammation and tissue tension are reduced postoperatively, this compartment block gradually resolves and the central flap whitening and macrofolds slowly fade. There is no clinically relevant necrosis or stromal remodeling. If the patient exhibits loss of BSCVA it is a result of unresolved macrofolds in the flap.

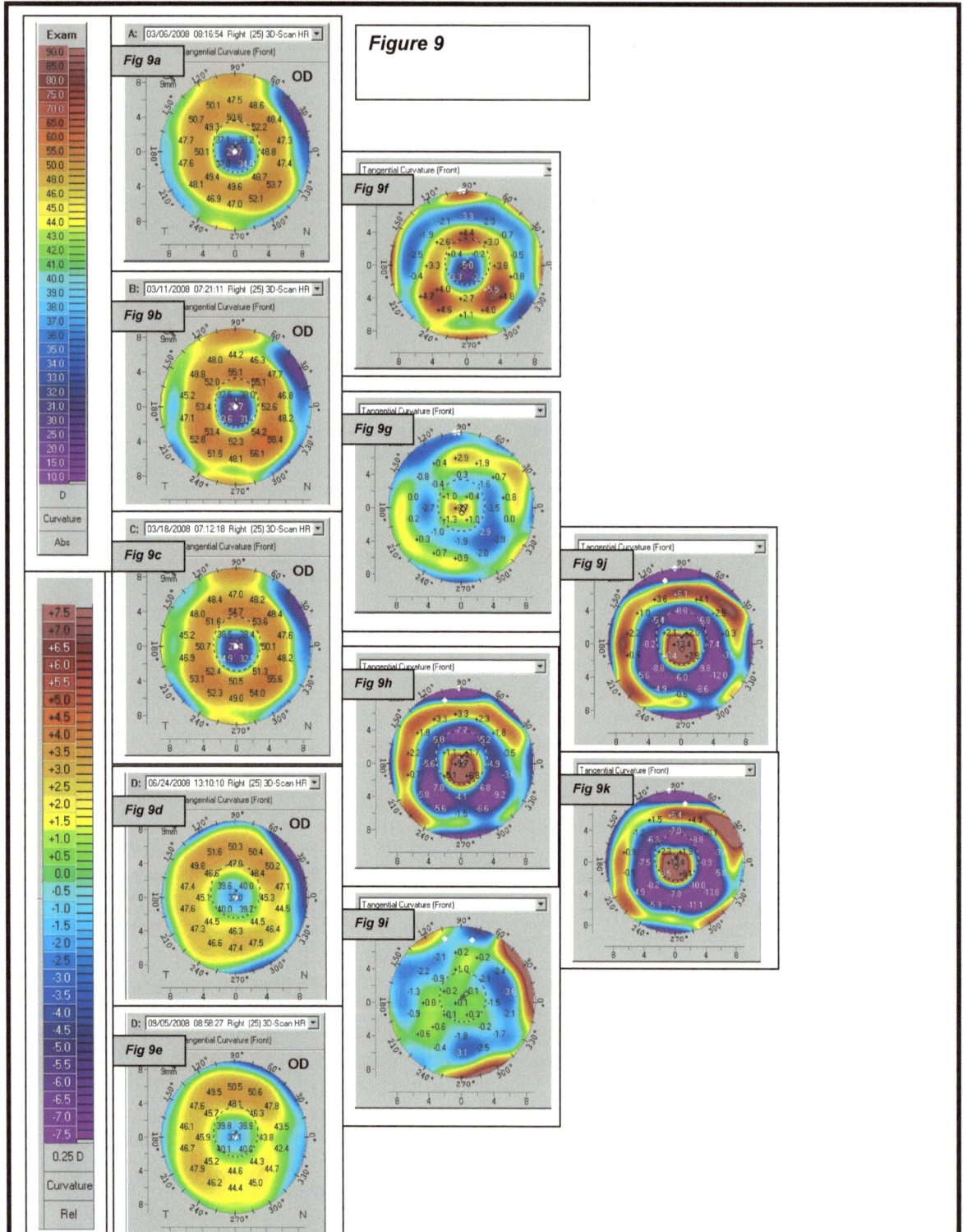

Figure 9 - Corneal power change caused by Severe KME

A hallmark characteristic of KME is a marked hyperopic shift currently attributed to stromal necrosis. However, sequential Pentacam™ tangential power maps demonstrate that the hyperopic shift occurs as a result of a very large change in corneal curvature. Moreover, this change is rapidly reversible. The pattern of events depicted by Pentacam™ tomography and

OCT imaging demonstrates a progressive process of central corneal flattening, mid-peripheral steepening and peripheral flattening that occurs within the first 10 days following presentation of severe KME followed by central corneal steepening, mid-peripheral flattening and peripheral steepening over the next few weeks. This reversible change in corneal shape causes the cornea to experience a marked hyperopic shift followed by a marked myopic shift. Our model posits that these events are governed by reversible biomechanical forces modulated by Pif, tissue tension and tissue compliance. Inflammatory cytokines, inside-out transmembrane signaling, pharmacologic agents and tissue toxins directly affect control of Pif.

Figure 9a through 9e depict sequential tangential power maps for a patient with severe KME. The time intervals for these maps are T=0, T=5 days, T=12 days, T=16 weeks and T=26 weeks respectively, where T=0 is the day of presentation of severe KME. These Pentacam™ images also correspond in time to OCT and Pentacam™ elevation images in Figure 10 and Figure 11. Scaling for images 9a–e is located on the upper left and scaling for images 9f-k is located on the lower left.

Images 9a through 9b demonstrate marked central corneal flattening, mid-peripheral steepening, peripheral flattening and a hyperopic shift. The change map 9f (9b-9a) suggests that over an interval of only 5 days, the central flattening is approximately – 5 D combined with approximately + 3.5 to + 5.5 D of peripheral steepening. This produces a marked hyperopic shift and a highly multifocal cornea.

Conversely, images 9c through 9e demonstrate marked central corneal steepening, mid-peripheral flattening, peripheral steepening and a very large compensatory myopic shift. The change map 9g (9c-9b) shows that over a subsequent interval of only 7 days from the nadir of the hyperopic shift, there is central steeping of approximately + 3.7 D and mid-peripheral flattening of approximately – 2D; change map 9h (9d-9c) shows that over the following interval of 14 weeks an additional + 9.7 D of central steepening and mid-peripheral flattening of approximately – 6 D occurs and; change map 9i (9e-9d) shows that over the final interval of 10 weeks an additional + 0.2 D of central steepening and mid-peripheral flattening of between – 1 and – 2 D occurs. Overall, this second stage of the process wherein corneal curvature is changed demonstrates a marked myopic shift that reverses the multifocality and the hyperopic shift that occurred initially over the first few days following presentation of severe KME.

Change map image 9j (9d-9b) attempts to summarize the healing process and demonstrates the overall amount of myopic shift that occurred during the 15 week period following the nadir of the hyperopic shift to be approximately + 13.4 D of central steepening and – 7 to – 8 D of mid-peripheral flattening. Similarly, change map image 9k (9e-9b) demonstrates the overall amount of myopic shift that occurred during the 25 week period following the nadir of the hyperopic shift to be approximately + 13.4 D of central steepening and – 7 to – 9 D of peripheral flattening. Maps 9j and 9k indicate that most, if not all of the significant recovery in this cornea following hyperopic shift from severe KME had occurred within 4 months of presentation. In this case, at 15 weeks, the patient's BSCVA was also 20/20 and the cornea was stable.

We believe that a comparison of these Pentacam™ change maps to time matched OCT images (Figure 12) demonstrates that this highly reversible marked change in corneal shape that occurs during KME is not caused by epithelial remodeling, stromal necrosis or stromal healing. Perhaps most compelling is the observation that 9f and 9h are nearly exactly mirror images of each other.

We posit that these tangential power maps demonstrate a biologically mediated reversible power change in the cornea that first creates a marked hyperopic shift followed immediately by a compensatory myopic shift. The clinical cause of this power change is marked central corneal flattening, mid-peripheral steepening and peripheral flattening followed by the reverse process of marked central steepening, mid-peripheral flattening and peripheral steepening. We further propose that this highly reversible and predictable process occurs as a result of loss of control of Pif followed by a reordering of corneal fluid dynamics, tissue tension and tissue compliance.

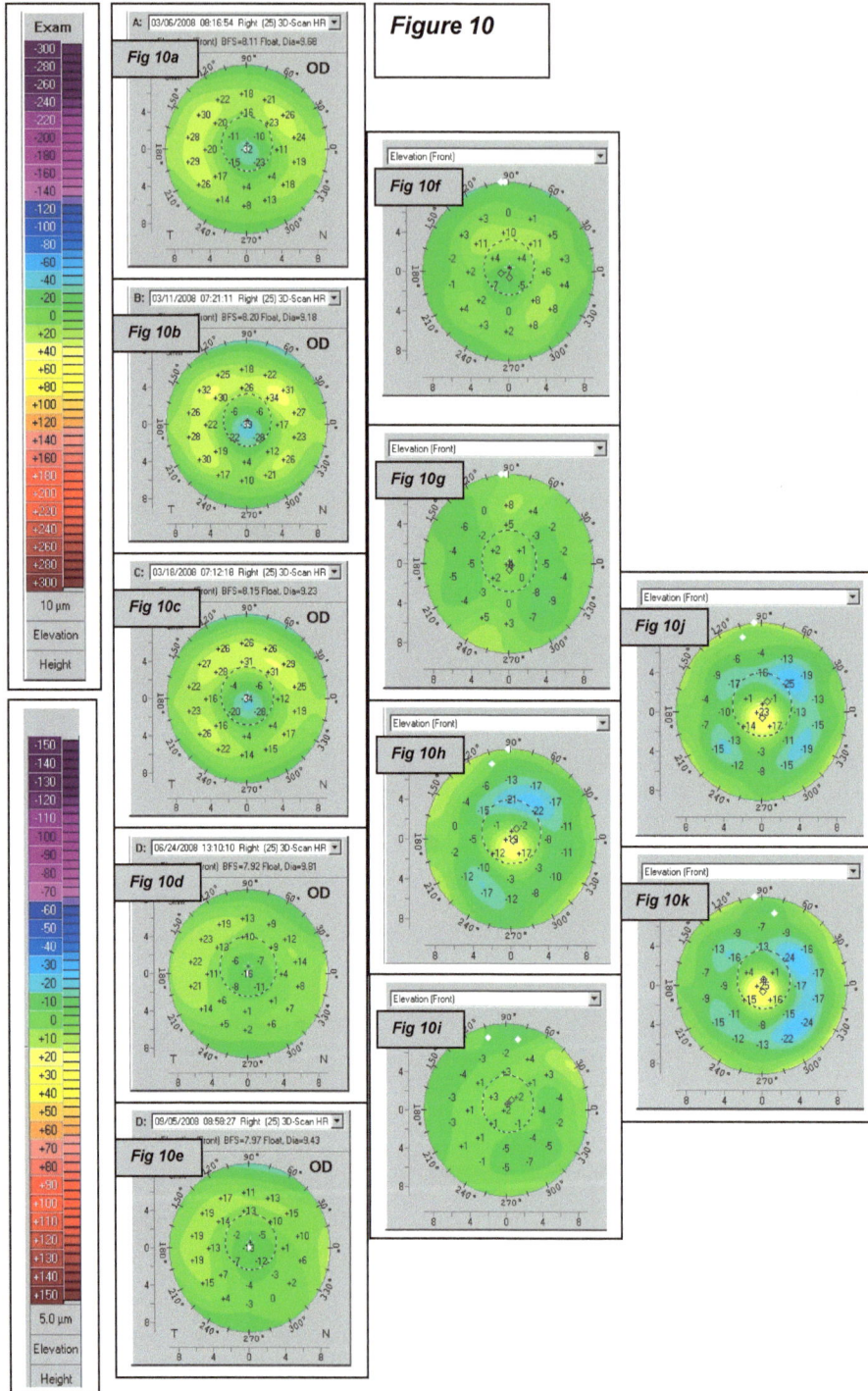

Figure 10 - Corneal elevation / shape change in Severe KME

Characteristic of KME is a marked hyperopic shift currently attributed to stromal necrosis. However, sequential Pentacam™ elevation maps demonstrate that the hyperopic shift occurs as a result of a very large change in corneal curvature. Moreover, this change is rapidly reversible. The pattern of events depicted by Pentacam tomography and OCT imaging demonstrates a progressive process of central corneal flattening, mid-peripheral steepening and peripheral flattening that occurs within the first 10 days following presentation of severe KME followed by central corneal steepening, mid-peripheral flattening and peripheral steepening over the next few weeks. This reversible change in corneal shape causes the cornea to experience a marked hyperopic shift followed by a marked myopic shift. Our model posits that these events are governed by reversible biomechanical forces modulated by Pif, tissue tension and tissue compliance. Inflammatory cytokines, inside-out transmembrane signaling, pharmacologic agents and tissue toxins directly affect control of Pif.

Figure 10a through 10e depict sequential front surface elevation maps for a patient with severe KME. We recognize that in the absence of a constant reference radius of curvature between measurements, change maps generated by subtracting various images are not entirely clinically accurate. However, we believe that the difference maps demonstrate important clinical trends in corneal shape and front surface elevation. The time intervals for these maps are T=0, T=5 days, T=12 days, T=16 weeks and T=26 weeks respectively, where T=0 is the day of presentation of severe KME. These Pentacam™ images also correspond in time to OCT and Pentacam™ elevation images in Figure 10 and Figure 11. Scaling for images 10a–e is located on the upper left and scaling for images 10f-k is located on the lower left.

Images 10a through 10b demonstrate marked central corneal flattening with mid-peripheral elevation consistent with a hyperopic shift. The change map 10f (10b-10a) suggests that over an interval of only 5 days, the central flattening is approximately 7 microns combined with approximately 10 microns of peripheral elevation. This produces a marked hyperopic shift and a highly multifocal cornea.

Conversely, images 10c through 10e demonstrate marked central corneal elevations with mid-peripheral flattening and a very large compensatory myopic shift. The change map 10g (10c-10b) shows that over a subsequent interval of only 7 days from the nadir of the hyperopic shift, there is central steeping of approximately 2 microns and mid-peripheral flattening of approximately 5 microns; change map 10h (10d-10c) shows that over the following interval of 14 weeks an additional 19 microns of central steepening and mid-peripheral flattening of approximately 15 to 20 microns occurs and; change map 10i (10e-10d) shows that over the final interval of 10 weeks an additional 2 microns of central steepening and mid-peripheral flattening of 4 microns occurs. Overall, this second stage of the process wherein corneal elevation and curvature is changed demonstrates a marked myopic shift that reverses the multifocality and the hyperopic shift that occurred initially over the first few days following presentation of severe KME.

Change map image 10j (10d-10b) attempts to summarize the healing process and demonstrates the overall amount of myopic shift that occurred during the 15 week period following the nadir of the hyperopic shift to be approximately 23 microns of central steepening and 20 to 25 microns of peripheral flattening. Similarly, change map image 10k (10e-10b) demonstrates the overall amount of myopic shift that occurred during the 25-week period following the nadir of the hyperopic shift to be approximately 25 microns of central steepening and 20 to 25 microns of peripheral flattening. Maps 10j and 10k indicate that most, if not all of the significant recovery in this cornea following hyperopic shift from severe KME had occurred within 4 months of presentation. In this case, at 15 weeks, the patient's BSCVA was also 20/20 and the cornea was stable.

We believe that a comparison of these Pentacam™ change maps to time matched OCT images (Figure 12) demonstrates that this highly reversible marked change in corneal shape that occurs during KME is not caused by epithelial remodeling, stromal necrosis or stromal healing. Perhaps most compelling is the observation that 10f and 10h are nearly exactly mirror images of each other.

We posit that these elevation maps demonstrate a biologically mediated reversible power change in the cornea that first creates a marked hyperopic shift followed immediately by a compensatory myopic shift. The clinical cause of this power change is marked central corneal flattening and mid-peripheral steepening followed by the reverse process of marked central steepening and mid-peripheral flattening. We further propose that this highly reversible and predictable process occurs as a result of loss of control of Pif followed by a reordering of corneal fluid dynamics, tissue tension and tissue compliance.

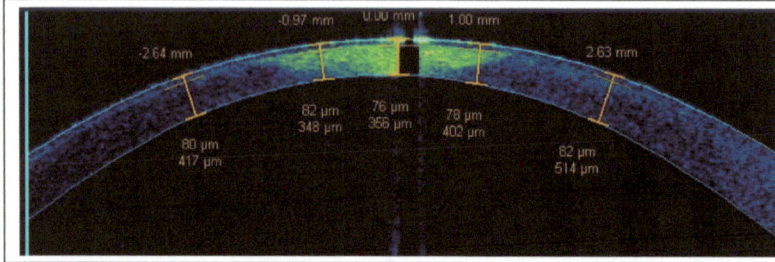

Figure 11a
T = 0

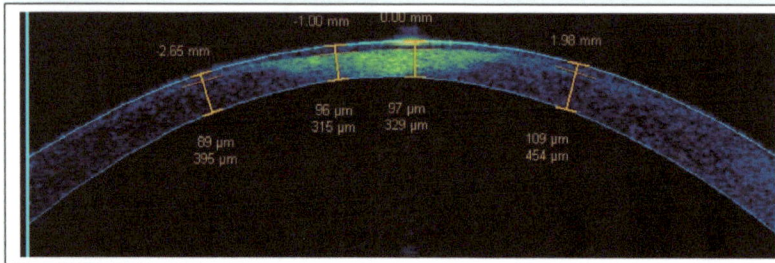

Figure 11b
T = Day 5

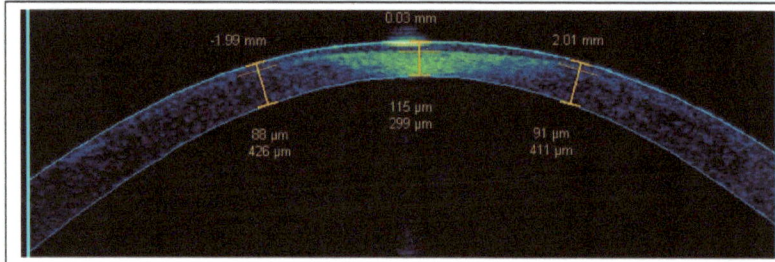

Figure 11c
T = Day 12

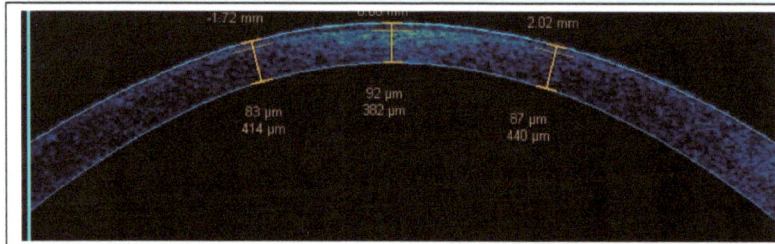

Figure 11d
T = Week 16

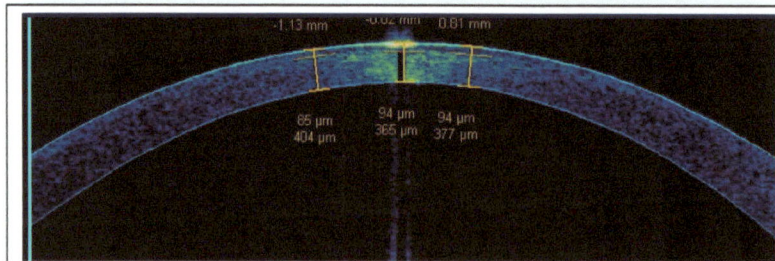

Figure 11e
T = Week 26

Figure 11 - Optical Coherence Tomography Imaging in Severe KME

In severe KME, under slit lamp examination the flap appears white and the posterior stromal bed appears thin. OCT imaging of the cornea demonstrates changes in the cornea that explain these findings. These images also assist in understanding the cause for the initial central corneal flattening and mid-peripheral steepening that is immediately followed by central corneal steepening and mid-peripheral flattening.

Figure 11a through 11c show marked flap thickening, particularly in the center of the flap, consistent with stromal edema. Accompanying this flap thickening is a progressive marked thinning of the central residual stromal bed combined with mid-peripheral stromal thickening. As the cornea recovers over subsequent weeks (figure 11c through 11e), the flap thickening resolves completely and the central thinning and mid-peripheral thickening are also markedly reversed.

OCT images are consistent with our model wherein we postulate that central corneal flattening is caused by mid-peripheral corneal thickening that creates biomechanical stress. Likewise, central flap edema is caused by a severe decrease in tissue compliance in the central residual stromal bed that causes a compartment block. The inability of fluid to move freely from the flap to the central stroma results in fluid accumulation in the flap. These events are reversible wherein a decrease in mid-peripheral stromal thickening reduces biomechanical stress and results in central corneal steepening. Reduced tissue tension improves the tissue compliance of the central stromal and allows the compartment block to clear over time. There is no clinically relevant stromal necrosis.

We believe that a comparison of these OCT images to time matched Pentacam™ change maps (Figures 9 and 10) demonstrates that this highly reversible marked change in corneal shape is not caused by epithelial remodeling, stromal necrosis or stromal healing.

We posit that these OCT images are consistent with our model for KME wherein we suggest that these clinical effects are created by a loss of control of Pif with resulting changes in corneal fluidics, local tissue hydration, corneal tension and tissue compliance. These latter changes result in a biologically mediated

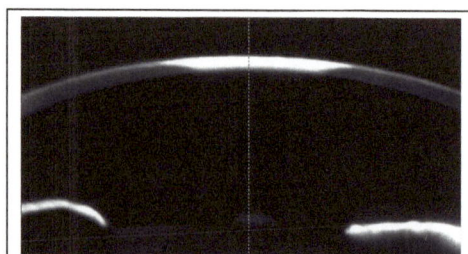

Figure 12a

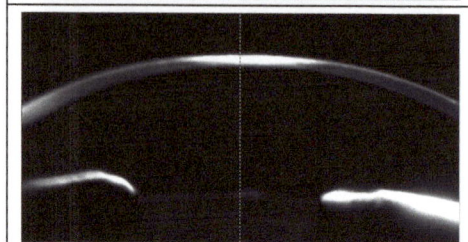

Figure 12b

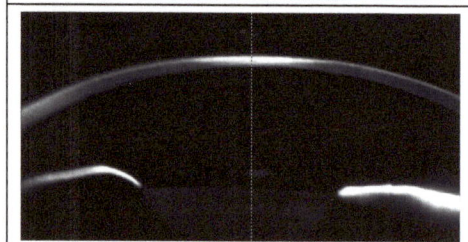

Figure 12c

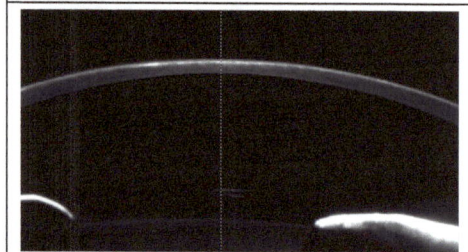

Figure 12d

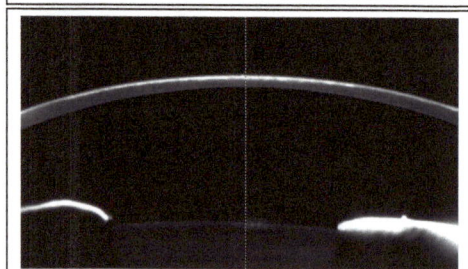

Figure 12e

Figure 12

Scheimpflug camera images from the Pentacam™ demonstrate the "cone shaped" white opacity extending from below the epithelium through the flap. Although we suggest that these images demonstrate some shadow artifact caused by intense light scattering by focal central flap edema that makes the cone appear to extend deeply in the residual stromal bed, we believe that these images nevertheless provide some insight into fluidics in the cornea.

Specifically, we posit that the movement of bulk fluid flowing from limbus to cornea center follows an inverse parabolic trajectory created by the anisotropic structural properties of the corneal ECM and the unique fluidic properties of proteoglycans subjected to tension and compression forces. These factors, combined with a compartment block created by loss of central stromal tissue compliance due to increased collagen tension, contribute to marked fluid accumulation in the central anterior cornea and flap. Clinically, this marked tissue edema creates an intense whitening of the central flap due to light scattering observed at the slit lamp and central flap macrofolds.

We posit that these Pentacam™ images demonstrate the anisotropic fluidic patterns characteristic of the cornea and document the compartment block that occurs in the anterior central stroma.

The Physiologic Role of the Corneal Stroma in Maintaining Corneal Hydration Homeostasis

To understand our model as well as the basic and clinical science supporting it, a review of our current understanding of those factors determining the hydration status of the cornea is necessary. Included in this discussion are the following core concepts.

> The model requires a solid understanding of the following:
>
> 1. How swelling pressures are regulated in interstitial tissue
> 2. The non-homogenous distribution of interstitial fluid pressures in the corneal stroma
> 3. Fluid dynamics in the corneal stroma
> 4. The role of the ECM and keratocyte in regulating interstitial fluid pressure
> 5. The impact of mechanical tissue stress, inflammatory cytokines and chemokines, pharmacologic agents and toxins in the regulation of interstitial fluid pressures in the cornea

General Concepts

The mechanism by which the cornea maintains its hydration and transparency has been generally considered to be well understood[101-103]. In its classical description, the corneal stroma consists of collagen fibers and keratocytes surrounded by an amorphous ground substance made up of highly negatively charged glycosaminoglycans. This biological matrix is very hydrophilic and provides a driving force for fluid accumulation and edema under pathologic conditions. The role assigned to the control of hydration of this stromal matrix has been entirely passive, with the belief that swelling pressure in the stroma was the product of strong but static osmotic forces created entirely by the compression of glycosaminoglycans.

In addition, it is well accepted that ion permeabilities and ion transport mechanisms provide the necessary conditions to maintain normal corneal hydration in the face of this tendency of the stroma to imbibe fluid from the tear film and aqueous. Epithelial or endothelial injury results in fluid influx into the stroma that may result in gross corneal swelling and loss of transparency if the compromise in permeability and or

> We believe that this traditional view of control of fluid dynamics in the cornea has been significantly challenged by our observations of the cornea's ability to respond in highly predictable manner to surgical injury, particularly in the context of refractive surgery

ion transport function is severe enough. A significant body of research underscores the complexity of these ion pumps and barriers to fluid movement. However, irrespective of this widely held paradigm, the particular local anatomic and physiologic changes in the stroma that occur during this swelling process are actually not well understood. In addition, structures such as Bowman's membrane are considered to be important anatomic and structural elements in the cornea but their potential role in the control of fluid flux is not widely appreciated.

We believe that this traditional view of control of fluid dynamics in the cornea has been significantly challenged by our observations of the cornea's ability to respond in highly predictable manner to surgical injury, particularly in the context of refractive surgery. We further believe that attempts to understand the mechanisms driving complications of refractive surgery such as DLK, Stage 4 DLK, CTK, CFN, FNS, CLK and epithelial ingrowth have further challenged this traditional view. In order to understand tissue behaviors that appear to deviate significantly from this classical view, we have incorporated into our model features that we believe describe a more comprehensive view of the control of corneal fluid dynamics than has been previously considered.

The Control of Interstitial Fluid Pressures

In general, physiologic systems that maintain structures or processes within strict parameters in the human body include complex feedback loops, systematic redundancy in organization and a broad distribution of control that incorporates multiple layers of process regulation. Our traditional view of the control of corneal hydration lacks virtually all of these characteristics. For example, with respect to maintenance of a negative Pif, the corneal epithelium, stroma and endothelium are currently perceived as operating independently. Provided the observed stability of corneal shape and visual acuity, it is our position that a model lacking such characteristics is highly unlikely to describe the complexity needed to deliver quality vision under the remarkable diversity of physiologic environments encountered in everyday life.

Physiologic systems that maintain structures or processes within strict parameters in the human body include complex feedback loops, redundancy and a broad distribution of control. Our traditional view of the control of corneal hydration lacks virtually all of these characteristics. With respect to maintenance of a negative Pif, the corneal epithelium, stroma and endothelium are currently perceived as operating independently. This is simplistic and naive.

As defined by the traditional fluid dynamics paradigm, a healthy endothelial pump combined with a relatively impermeable epithelial layer and endothelial tight junctions are essential elements in the creation of a negative Pif. What remains unclear however, is precisely how such a system controls and regulates Pif under the wide dynamic range of conditions experienced by the cornea.

To better understand the difference between creation of pressure and control or regulation of pressure via a physiologic feedback loop, a simple analogy appears useful. For example, the human heart is responsible for the creation of blood pressure. Certainly, without the heart there is no blood pressure. Similarly, without the endothelial pump there is no negative Pif. However, the fact that the heart is essential to blood pressure does not provide guidance with respect to how blood pressure is controlled in the human body. It would be unlikely to find a physiologist today that held the simplistic view that blood pressure is controlled entirely by the heart. Blood pressure regulation is highly complex and incorporates multiple feedback loops including the participation of other organ systems, hormones, vascular tone, interstitial fluid pressures, electrophysiology and specific cell functions.

> It is our position that to understand the impact of corneal refractive surgery on local tissue physiology, we must begin to recognize that our current paradigm wherein the stroma plays a passive role in tissue hydration dynamics is also too simplistic

It is our position that to understand the impact of corneal refractive surgery on local tissue physiology, we must begin to recognize that our current paradigm wherein the stroma plays a passive role in tissue hydration dynamics is also too simplistic. Moreover, it is our view that we must clearly differentiate between the mechanics of those factors that control or regulate Pif versus those that create Pif. Unfortunately, in our review of the scientific literature, we have been unable to identify a model that describes a feedback loop in the cornea that would control or regulate Pif in the face of changing physiologic conditions or corneal refractive surgery. In addition, there does not appear to be a model that suggests how a Pif gradient in the cornea, wherein Pif varies depending upon specific tissue localization, can either be created or maintained.

> We must clearly differentiate between the mechanics of those factors that control or regulate Pif versus those that create Pif

In the absence of an understanding of these Pif control mechanisms, we propose a belief that current concepts regarding the control of corneal transparency and shape are incomplete and inaccurate. Specifically;

1. With respect to normal corneal fluid dynamics, the cornea is much more of an "open" system than the "closed" system controlled by tight junctions as has been traditionally described. For example, there appears to be no specific barrier to fluid movement from the limbus into the stroma.

2. Particularly during and after excimer laser refractive surgery, a passive stroma should imbibe fluid uncontrollably thereby producing highly unpredictable corneal shape and clarity during the first few days following either PRK, LASEK or

LASIK surgery. Other than under exceptional circumstances, this does not occur.

3. The shape and clarity of the cornea remains constant irrespective of a broad spectrum of environmental and physiologic conditions to which the cornea is subjected

4. Current ideas fail to explain the stability of the cornea with respect to shape and clarity despite

   a. Significant variations in the number and health of endothelial cells present in a specific cornea

   b. Marked variations in the thickness of the stroma in the normal population

   c. The effects of tissue strain or tension on Pif

   d. The adverse effects of extreme physiologic conditions that would impact instantaneous endothelial function

In that context, we posit that it is naive to believe that a monolayer of endothelial cells can provide for a control system that can generate a reproducible corneal shape, a homeostatic Pif, a Pif gradient and a consistent level of corneal clarity particularly over the spectrum of physiologic conditions to which the

> It is naive to believe that a monolayer of endothelial cells can provide for a control system that can generate a reproducible corneal shape, a homeostatic Pif, a Pif gradient and a consistent level of corneal clarity particularly over the spectrum of physiologic conditions to which the cornea is subjected

cornea is subjected. The cornea's ability to maintain these functions as well as relatively constant refractive power is remarkable despite the disruptive effects of laser refractive surgery, nocturnal or contact lens induced hypoxia, marked changes in intraocular pressures during strenuous exercise, intraocular inflammation, variations in nutrient availability in the aqueous, sudden exposure to highly humid or arid environments or even epithelial defects from ocular disease, dry eye or minor trauma.

> The ability of the cornea to maintain a consistent shape and degree of tissue clarity in the context of such marked variations in epithelial or endothelial health and metabolic function suggests the need for dynamic control processes, feedback systems and an active stroma

It is our position that the ability of the cornea to maintain a consistent shape and degree of tissue clarity in the context of such marked variations in epithelial or endothelial health and metabolic function suggests the need for dynamic control processes, feedback systems and an active stroma. Current traditional concepts regarding the control of corneal hydration and fluid dynamics do not incorporate this level of complexity, systematization, adaptability or functionality.

Of particular concern is that the cornea is actually not a closed system controlled by an epithelial barrier and an endothelial pump, as proposed by classical fluid dynamic theory. Rather, fluid readily flows from the limbus and sclera into the cornea. Despite such fluid movement, the cornea is uniformly transparent. In a system with a passive stroma, it is difficult to devise a mechanism that will provide essentially uniform corneal transparency in the face of unchecked influx of limbal

> The cornea is actually not a closed system controlled by an epithelial barrier and an endothelial pump, as proposed by classical fluid dynamic theory

fluid. As a result, we believe that a comprehensive model of control of Pif needs to incorporate a mechanism that can maintain a uniform degree of corneal clarity from its periphery to its center in a non-closed system that is open to fluid inflow from the sclera and can adapt to fluid influx from the limbus in response to a local change in Pif.

> A comprehensive model of control of Pif needs to incorporate a mechanism that can maintain a uniform degree of corneal clarity from its periphery to its center in a non-closed system that is open to fluid inflow from the sclera and can adapt to fluid influx from the limbus in response to a local change in Pif

An additional challenge to current views on corneal fluid dynamics is an understanding of how a monolayer of endothelial cells creates a tissue effect hundreds of microns away. Intuitively, it seems reasonable to suggest that there is a distance from that monolayer (or any

endothelial cell layer in the body) where there is little if any physiologic effect. If so, the effect of an ion pump must decay in some manner with increasing separation of the target tissue from the ion pump. It is important to consider how this physiologic process occurs in both a classical static or "closed" system as well as in a highly dynamic situation as is found in refractive surgery. Additional complicating factors include a recognition that the cornea of an individual patient is markedly thicker at its periphery compared to the center and that the variation in corneal thickness within the general population is significant. Irrespective of these variations, corneal clarity and collagen periodicity is generally assumed to be uniform across the cornea in a given patient, is thought to be maintained within a narrow range in order to provide for tissue transparency within the human population of corneas as a whole and does not appear to change substantively with patient age.

The local effect of an ion pump on a target tissue is a basic concept that serves as a fundamental building block for any model of corneal fluid dynamics. First, we suggest that there is complex relationship between the location of an ion pump and its physical distance from a specific target tissue. Second, we propose that the health and physical status of that target tissue, including its Pif, may influence the local effectiveness and ability of that ion pump to control Pif or remove fluid from that local tissue. A third factor that we believe is important to local ion pump efficiency and effect is an understanding of the relationship between rates of fluid leakage or influx and its effect on local tissue Pif and hydration, particularly during traumatic loss of the epithelial barrier. If fluid leakage is located at a point distal to the location of the pump (as occurs in laser refractive surgery), Pif and the state of hydration of corneal tissue is likely no

> A system in which the degree of hydration is controlled exclusively by the existence of an intact epithelium and a normally functioning endothelium will markedly swell in their absence, even if temporary. This will allow a passive ECM to rapidly and aggressively imbibe fluid thereby affecting both tissue transparency and corneal shape

longer a homeostatic constant but rather represents a gradient that ranges from the point of maximum hydration adjacent to the point of fluid entry to a state of relative tissue dehydration located next to the pump. Unfortunately, the relationship between these factors is an area of basic and clinical science that has seen very little investigation. However, we believe that the relationship between an ion pump and its target tissue is very important to an understanding of the processes at work in refractive surgery and specifically in unraveling the forces involved in the pathogenesis of DLK, Stage 4 DLK, CTK, CFN, FNS, CLK and epithelial ingrowth.

We suggest that the efficiency of the endothelial layer decays as distance from the monolayer increases. However, in a truly closed system, we would postulate that it is likely that decay in pump efficiency with increasing distance to its target tissue is not of great importance.

This is because in a closed system with insignificant amounts of fluid inflow, tissue dehydration caused by continuous ion pumping can, over time, be expected to generate and sustain a steady hydration state and a homeostatic distribution of Pif.

However, with respect to fluid flux, the undisturbed cornea is not truly a closed system. Perhaps most importantly, a cornea subjected to the effects of PRK, LASIK or other corneal refractive procedures is clearly not a closed system. Despite such a significant physiologic challenge, which includes loss of barrier function and disruption of endothelial morphology, the cornea rather robustly maintains its clarity and shape.

We suggest that a system in which the degree of hydration is controlled exclusively by the existence of an intact epithelium and a normally functioning endothelium is unlikely to behave in such a manner since their absence, even if temporary, will allow a passive ECM to rapidly and aggressively imbibe fluid

> The failure of the cornea to respond in a manner consistent with the traditional fluidics model is of significant concern

thereby affecting both tissue transparency and corneal shape. Despite this rather obvious incongruity, the conventional paradigm for corneal fluid dynamics involves a tacit belief in a system wherein the hydration state of the stroma is exclusively controlled by closed system processes that demand an intact epithelium, a normally functioning monolayer of endothelial cells and a passive stroma as essential mechanisms to maintenance of corneal transparency and shape. We find the failure of the cornea to respond in a manner consistent with such a model to be of significant concern.

The Active Role of the Interstitium in Controlling Pif

In addition to considering what we perceive to be fundamental weaknesses in our understanding of how the cornea adapts to markedly altered physiologic conditions, we believe it to be both useful and instructive to consider how Pif is controlled in comparable tissue systems throughout the body. We suggest that such an analysis may provide insight into physiologic process that occur in interstitial tissues in general and provide a vision into how similar processes might occur in the cornea.

> We believe it is instructive to consider how Pif is controlled in comparable tissue systems throughout the body

Over the past several decades, the field of physiology has explored in detail how interstitial fluid pressure is controlled in numerous tissue systems. Unfortunately, the classical understanding of fluid dynamics and control of Pif in the cornea described previously in this paper[101-103] is markedly at odds with current knowledge with respect to Pif regulation in the rest of the body. In that context, we can either assert that the control of Pif in the cornea

is completely unique compared to all other tissue systems studied to date or we might conclude that our current understanding of how Pif is controlled in the cornea is highly flawed.

We posit that the control and regulation of Pif in the cornea abides by similar physiologic processes observed in tissue of comparable embryonic origin. In support of that opinion, we think that it is useful to explore current concepts in the

> We posit that the control and regulation of Pif in the cornea abides by similar physiologic processes observed in tissue of comparable embryonic origin. We think that it is useful to explore current concepts in the physiology of Pif control, as the field of Ophthalmology has generally ignored these advancements

physiology of Pif control, as the field of Ophthalmology has generally ignored these advancements.

In recent years, the classical understanding on the subject of control of corneal interstitial swelling pressures has been significantly challenged by a larger body of experimental and clinical work in comparable tissues found through the body. A recent review article summarizes these findings and suggests that a more active role needs to be placed on the connective tissue cells and extracellular matrix in regulating Pif[55].

Meyer[104] observed that under normal conditions, the tendency for connective tissue to expand is counteracted by collagen and microfibril networks physically restraining the swelling hyaluronan. This observation was pivotal towards the formulation of an understanding of the fundamental processes that control fluid dynamics. We now understand that interstitial cells contained in connective tissue exert active control over tissue swelling in inflammation and in the development of

> We now understand that interstitial cells contained in connective tissue exert active control over tissue swelling in inflammation and in the development of edema following tissue injury. Interstitial cells have also been demonstrated to respond to inflammatory cytokines, hormonal and pharmacologic intervention and that these control mechanisms serve to either induce or reverse tissue edema

edema following tissue injury. Moreover, β_1–integrin adhesion receptors provide a common pathway by which these cells can raise as well as lower interstitial fluid pressure. Interstitial cells have also been demonstrated to respond to inflammatory cytokines, hormonal and pharmacologic intervention and that these control mechanisms serve to either induce or reverse tissue edema.

Lund et al.[105] measured interstitial fluid pressure (Pif) in full thickness, burn injury in the rat. They found, somewhat surprisingly, that Pif did not rise to positive values during edema generation as previously thought, but actually decreased from –1 to –150 mmHg. These observations have been subsequently confirmed by others[106-109]. It is this marked negativity in Pif that explains the rise in net capillary filtration. Lund's observations were the first to assign an active role to the interstitium in creating transcapillary fluid flux. Lund et al.[105] hypothesized that there were three possible explanations for the change in Pif: 1) water loss from evaporation; 2) generation and immobilization of new colloids attributed to heat denaturation of collagen to gelatin; and 3) the creation of expansile forces in the interstitium by physical changes in the biomatrix. Subsequent studies have implicated the third explanation as the primary mechanism governing fluid dynamics in acute burn injury.

Similar to burn injury, acute inflammation is also associated clinically with the rapid formation of tissue edema. During acute inflammation, numerous studies have demonstrated an increased negativity of Pif in the inflamed tissue. For example, Koller and Reed[110] demonstrated that in the anaphylactic response to dextran which results in edema formation in rats, Pif is also markedly reduced, although it remains closer to normal physiologic levels than in 2nd or 3rd degree burns. Koller et al.[111] showed that mast cell degranulation markedly decreased Pif, thereby further demonstrating that the interstitium is an active participant in the edema generating process and suggests an interaction between the structural and regulatory mechanisms of the interstitium and the hormonal or biochemical cascade surrounding mast cell degranulation.

Neurogenic inflammation, an inflammatory reaction characterized by edema formation and plasma protein extravasation in response to stimulation of the sensory fibers of the vagal nerve, also demonstrated a marked reduction in Pif [112]. The reduction in Pif started within 30 seconds after onset of stimulation and reached a steady state within minutes. Moreover, recent studies have suggested a complicated interplay between the sensory nerves, mast cells and loose connective tissue, suggesting an important role for the mast cell in eliciting at least some of the biological responses induced by stimulation of sensory nerves[113].

Mechanisms involved in edema formation – the role of β_1–integrins:

The observation that lowering the Pif is a general mechanism participating in the development of inflammatory edema is only one aspect of understanding this phenomenon. An examination of the underlying biological and molecular events surrounding this process is instructive in understanding the events that likely occur in the human cornea.

Transmembrane proteins known as integrins play a pivotal role in control of Pif and edema formation. They perform three primary functions that include:

1. Sensing tension in the ECM
2. Creating tension between the interstitial cell and the ECM
3. Transmitting information from the ECM to the cell (outside-in transmembrane signaling)

Transmembrane proteins known as integrins play a pivotal role in control of Pif and edema formation. They perform three primary functions that include:

1. Sensing tension in the ECM
2. Creating tension between the interstitial cell and the ECM
3. Transmitting information from the ECM to the cell (outside-in transmembrane signaling)

In the control of Pif, the most important role for integrins is to function as an integral element to the transmission of force or tension between the cell and its environment. However, an additional important role for integral membrane proteins is to transfer information across the plasma membrane, a process known as transmembrane signaling. This process allows the interstitial cell to sense tension and adapt to its external environment. Several cell-adhesion molecules have been demonstrated to carry out this function. Integrins and cadherins can transmit signals from the extracellular environment to the cytoplasm by means of linkages with the cytoskeleton and with cytosolic regulatory molecules, such as protein kinases and G proteins. Protein kinases activate or inhibit their target proteins through phosphorylation, whereas G proteins activate or inhibit their protein targets through physical interaction. The engagement of an integrin with its ligand can induce a variety of responses within a cell, including changes in cytoplasmic pH or Ca^{2+} concentration, protein

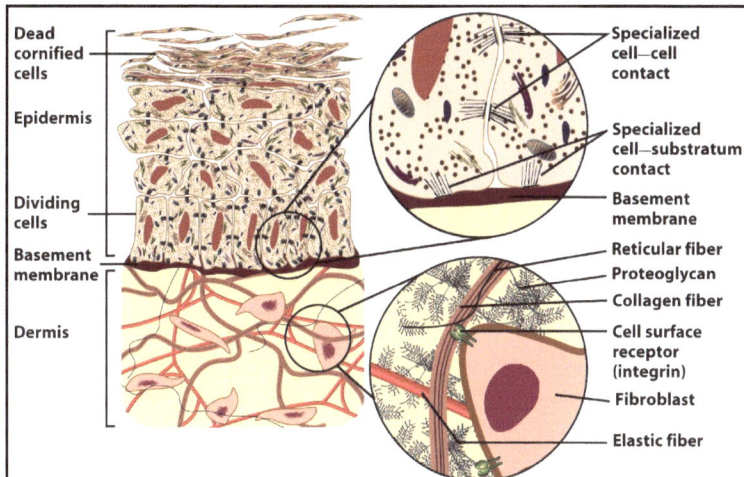

Dead cornified cells
Epidermis
Dividing cells
Basement membrane
Dermis

Specialized cell—cell contact
Specialized cell—substratum contact
Basement membrane
Reticular fiber
Proteoglycan
Collagen fiber
Cell surface receptor (integrin)
Fibroblast
Elastic fiber

Figure 13 - Role of Integrins In the control of Pif, the most important role for integrins is to function as an integral element to the transmission of force or tension between the cell and its environment. However, an additional important role for integral membrane proteins is to transfer information across the plasma membrane, a process known as transmembrane signaling. This process allows the interstitial cell to sense tension and adapt to its external environment.

associated α- and β- subunits. Integrins mediate intracellular as well as cell-extracellular matrix adhesion[115-117]. Four integrins are known to bind triple helical interstitial collagens, namely the $\alpha_1\beta_1$, $\alpha_2\beta_1$, $\alpha_{10}\beta_1$ and $\alpha_{11}\beta_1$ integrins[117,118].

Integrins, or their complexes, function as mechano-receptors, which sense tension exerted on cells by extracellular matrix structures[119-121]. Integrins have been further demonstrated to transmit signals from the external environment to the cell interior, a phenomenon known as "outside-in" transmembrane signaling. The binding of the extracellular domain of an integrin to a ligand, such as fibronectin or collagen, can induce a conformational change at the

phosphorylation and gene expression. These changes, in turn, can alter a cell's interaction with the extracellular matrix, locomotion, migratory activity, growth potential, state of differentiation or viability[114].

The role of β_1–integrins in the development of inflammatory edema has been explored by the use of cell-mediated contraction of three-dimensional gels in vitro in parallel with in vivo experiments.

Integrins are transmembrane heterodimeric glycoproteins composed of non-covalently

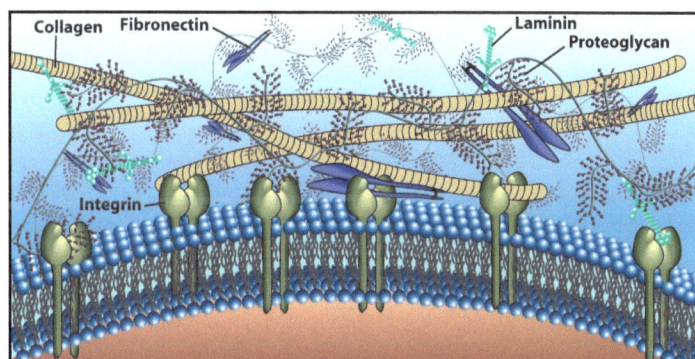

Collagen Fibronectin Laminin Proteoglycan Integrin

Figure 14 - Organization of the ECM
Collagen, fibronectin and laminin demonstrate binding sites for each other as well as integrins located in the cell membrane.

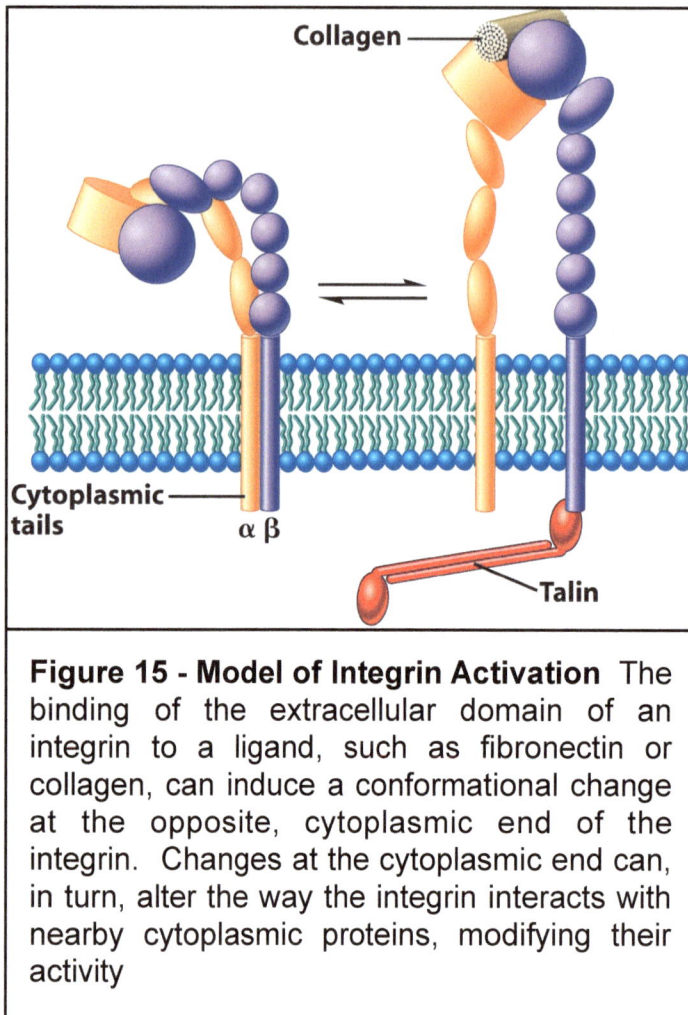

Figure 15 - Model of Integrin Activation The binding of the extracellular domain of an integrin to a ligand, such as fibronectin or collagen, can induce a conformational change at the opposite, cytoplasmic end of the integrin. Changes at the cytoplasmic end can, in turn, alter the way the integrin interacts with nearby cytoplasmic proteins, modifying their activity

opposite, cytoplasmic end of the integrin. Changes at the cytoplasmic end can, in turn, alter the way the integrin interacts with nearby cytoplasmic proteins, modifying their activity. Thus, as integrins either actively bind or uncouple from an extracellular ligand, they can trigger the activation of cytoplasmic protein kinases, such as focal adhesion kinase (FAK) or Src. These kinases can then phosphorylate other proteins initiating a chain reaction. Such outside-in signals transmitted by integrins can influence many aspects of cell behavior including motility, growth and apoptosis[114].

In addition to conveying mechano-receptors signals, integrins are active in force transmission from the cell interior to the extracellular matrix. This has been demonstrated in numerous studies[122]. Fibroblasts cultured in a three-dimensional collagen lattice contract the lattice using traction forces generated by the cell culture[123-126]. In addition, the contraction process is stimulated by platelet-derived growth factor isoform BB (PDGF-BB)[127,128] and depends upon β_1 integrins[127], particularly the collagen-binding integrin $\alpha_2\beta_1$[129,130]. Conversely, the addition of antibodies specific to β_1 integrins can slow or attenuate the contraction[127]. When the antibody is removed from the medium, the contraction process resumes thereby demonstrating a completely reversible cellularly mediated process dependent upon the function of β_1 integrins.

Mechanisms involved in edema formation – the role of the cellular cytoskeleton

In vitro studies using three dimensional collagen gels have provided substantial evidence implicating the role of the cellular cytoskeleton in the mediation of contractile forces in the biomatrix of the interstitium[127,131]. Cytochalasin D, which

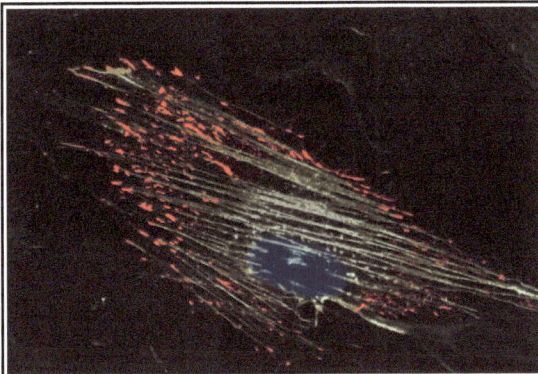

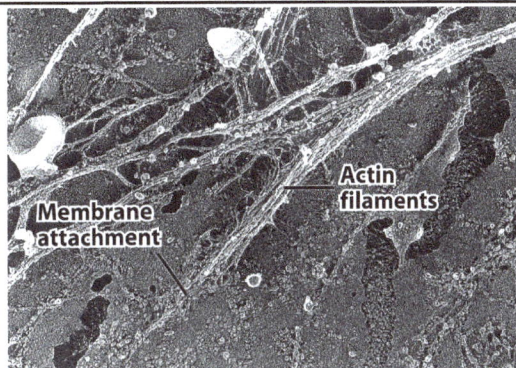

Figure 16 - Structure of Focal Adhesions (a) the cell has been stained with fluorescent antibodies that show the locations of the integrins (red) and the actin filaments (gray-green). Integrins are located in patches that correspond to focal adhesions. (b) a cell processed by quick-freeze, deep-etch analysis demonstrates a focal adhesion. Bundles of actin filaments associate with the inner surface of the cytoplasmic membrane in the region of a focal adhesion.

Figure 17 - Mechanism of Outside In Transmembrane Signaling Drawing of a focal adhesion demonstrating the interaction of integrins with proteins on both sides of the phospholipid membrane. The binding of extracellular ligands such as collagen and fibronectin is though to produce conformational changes in the cytoplasmic domains of the integrins which cuase the integrins to become linked to actin filaments of the cytoskeleton. Linkages with the cytoskeleton are mediated by vairous actin-binding proteins, such as talin and α-actin, that bind to the β subunit of the integrin. The cytoplasmic domains of integrins are also associated with protein kinases, such as FAK (focal adhesion kinase) and Src. The attachment of the integrin to

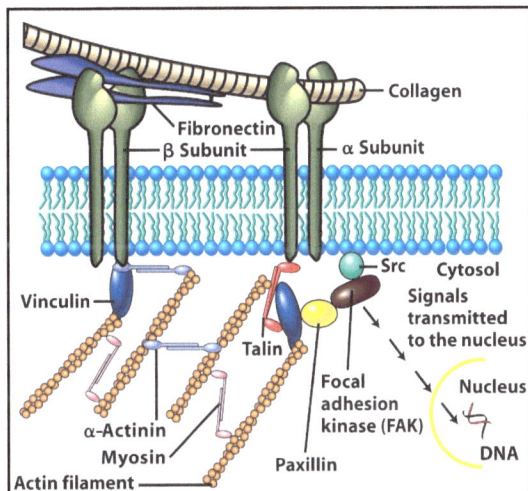

an extracellular ligand can active these protein kinases and start a chain reaction that transmits signals thoughout the cell. Such outside-in signals transmitted by integrins can influence many aspects of cell behavior including motility, growth and apoptosis

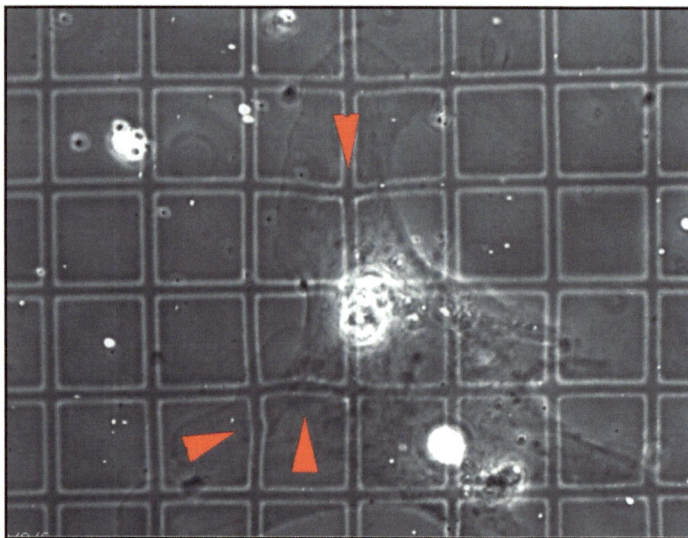

Figure 18 - Contraction forces exerted by focal adhesions Fibroblasts plated on a deformable grid create traction and deformation of the matrix. Contraction force exerted by actin cytoskeleton is transmitted to the matrix via focal adhesions whose principle component include integrins. It is this contractile force, modulated by inflammatory cytokines and other messengers that control Pif

disrupts the actinomyosin contractile apparatus, blocks three-dimensional gel contraction[132]. In addition, the rate of gel contraction is affected by several growth factors and cytokines known to affect actin myofilament function. Platelet-derived growth factor (PDGF) and endothelin have been demonstrated to enhance the contraction process. Conversely, interleukin-1 and TNF-α have been shown to slow the rate of contraction. Increased levels of the intracellular second messenger cAMP will inhibit the contraction process, suggesting that it plays a role in process control and modulation. Each of these observations implicate the actinomysin contractile apparatus of the cell as the factor that provides the interstitial connective tissue cells with the contractile force, which when reversed or decreased would allow the tissue to swell.

These in vitro studies have been supported by in vivo studies. Reed et al.[133] used the polyclonal anti-β_1-integrin antibody to inhibit fibroblast mediated collagen adhesion in the interstitial tissue of rat paw. Treatment with this specific anti-β_1 integrin antibody lowered P_{if} and caused edema formation.

The specific role of the cellular cytoskeleton in the process of controlling Pif via its intracellular actin filament system has been demonstrated in several experiments. Berg et al.[134] showed that disruption of the F-actin in the cell cytoskeleton using the agent cytochalasin D generated a negative Pif in a dose-dependent manner. Moreover, Bronstad et al.[135], examined the

Cell cytoskeleton tension transmitted to the ECM through integrins can modulate Pif and can induce the formation of edema. The control cascade includes mediators of inflammation such as prostaglandins, interleukins and TNF-α and is effected by the intracellular second messenger cAMP. Factors that serve to elevate intracellular cAMP and mobilize intracellular Ca^{2+} affect the internal cell cytoskeleton and in turn the β-integrin mediated cell-collagen biomatrix interactions that control interstitial fluid dynamics effects of pretreatment with phalloidin, an agent that fixes the actin filaments within the cell, on the early phase of rapid edema in the dextran-induced anaphylaxis in the rat. Their study demonstrated that pretreatment with phalloidin, completely abolished the lowering of Pif and edema formation created by dextran injection. Clearly, intracellular actin plays a pivotal role in the creation of contractile forces affecting swelling pressure in interstitial tissues.

In summary, cell cytoskeleton tension transmitted to the ECM through integrins can modulate Pif and can induce the formation of edema. The control cascade includes mediators of inflammation such as prostaglandins, interleukins and TNF-α and is effected by the intracellular second messenger cAMP. Factors that serve to elevate intracellular cAMP and mobilize intracellular Ca^{2+} affect the internal cell cytoskeleton and in turn the β-integrin mediated cell-collagen biomatrix interactions that control interstitial fluid dynamics.

The Effects of Prostaglandins, Inflammatory Cytokines and Intracellular cAMP levels on Interstitial Fluid Pressure:

Prostaglandins are ecosanoids synthesized from arachidonic acid and have been demonstrated to exhibit a role in nearly every step of the inflammatory response in epidermis. In tissues they act as local hormones that affect different G-protein linked receptors. Their biological response is determined by the action of intracellular second messengers such as cAMP.

> Prostaglandins demonstrate a role in nearly every step of the inflammatory response. They act as local hormones that affect different G-protein linked receptors whose biological response is determined by the action of intracellular second messengers such as cAMP. They induce a marked increased negativity in Pif and enhance edema formation by inhibiting fibroblast contractility and cell cytoskeleton function

Several inflammatory mediators such as prostaglandin PGE_1 and interleukin-1 have been demonstrated to inhibit contractions of fibroblast-mediated collagen gel contraction[136-138].

Berg et al[139] were the first to demonstrate the in vivo effect of prostanoids on Pif. They evaluated the effects of PGE_1, PGI_2, and $PGF_{2\alpha}$ (lantanoprost) on dextran-induced anaphylaxis in rats and collagen gel contraction assays. PGE_1 and PGI_2 are potent vasodilators and $PGF_{2\alpha}$ is a potent vasoconstrictor. In this study, they were able to demonstrate a specific, dose-dependent effect on the creation of a negative Pif and an associated edema formation by PGE_1 and PGI_2 in rats. Moreover, $PGF_{2\alpha}$ reversed the increased negativity of Pif created by the anaphylactic reaction to dextran. They concluded that PGE_1 demonstrated the ability to bind to both EP and IP type receptors and it is this receptor binding that initiates the inhibition of fibroblast mediated collagen contraction.

Recognizing that EP and IP receptors promote the intracellular production of cAMP through the stimulation of adenylate cyclase and that the elevation of intracellular cAMP is known to inhibit the fibroblast mediated contraction of collagen gels, it is highly likely that cAMP is the second messenger in this control cascade. Moreover, their work demonstrates that the effects of PGE_1 and PGI_2 do not directly involve alterations in the functions of the collagen-binding β_1-integrins but work higher up in the control cascade. The most likely candidate is the contractile machinery of the cell and it has been shown in multiple cell types that cAMP exerts an effect on both actin and myosin light-chain kinase. Such an effect by cAMP could easily explain the effects of PGE_1 leading to the inhibition of collagen gel contraction.

> Clinical in vivo studies demonstrate that prostaglandins and other inflammatory cytokines applied locally significantly alter Pif, typically driving it more negative and causing local edema.

In summary, prostaglandins appear to have the capacity to induce a marked increased negativity in Pif and enhance edema formation by inhibiting fibroblast contractility and cell cytoskeleton function and this effect is likely mediated via EP and IP receptors and the elevation of cAMP in fibroblast cells.

In addition to basic science, clinical in vivo studies demonstrate similar effects by prostaglandins and other inflammatory cytokines. Salnikov et al.[140] demonstrated that the local injection of PGE_1 in experimental carcinomas significantly reduced Pif and markedly increased the delivery of low molecular weight cytostatic 5-FU in these tumors and measurably improved treatment efficiency and destruction of the carcinoma. They found that Pif reached a minimum within 10-15 minutes of PGE_1 application and returned to baseline within 60 minutes.

Rubin et al.[141] also demonstrated that the introduction of PGE_1 in the tissue of rat mammary carcinomas was able to significantly reduce Pif and enhance the transport of the chemotherapeutic drug 51Cr-EDTA into the tumor and enhance killing efficiency in solid malignancies.

Iversen and Reed[142] used a microdiaysis technique to evaluate the effects of PGE_1 on the interstitial space of rat skin. They found that PGE_1 induced the lowering of Pif and significantly improved transcapillary transport of 51Cr-EDTA into skin interstitium.

> Lipopolysaccharides (LPS), tumor necrosis factor-α, interleukin-6 and interleukin-1β all induce lowering of Pif.

Iversen et al.[143] also demonstrated increased plasma protein extravasation in rat and mouse skin most notably when exposed to PGE_1 and compound 48/80 (a mediator of mast cell degranulation).

Nedrebo et al.[144] tested the effects of lipopolysaccharides (LPS), tumor necrosis factor-α, interleukin-6 and interleukin-1β on Pif in a model of gram-negative sepsis. They found that LPS, tumor necrosis factor-α, interleukin-6 and interleukin-1β all induced lowering of Pif when given either intravenously or intra-arterially, whereas only tumor necrosis factor-α, interleukin-6 and interleukin-1β lowered Pif when given subdermally. They concluded that the tissue edema associated with gram-negative sepsis is related to the local tissue effects of tumor necrosis factor-α, interleukin-6 and interleukin-1β and that these factors served to antagonize cell-collagen biomatrix interactions that control Pif.

> Platelet activating factor (PAF) has been demonstrated to lower Pif likely through G-protein modulated pathways. Experimentally, DLK can be inhibited and perhaps prevented using PAF antagonists

Iversen et al.[145] investigated the effect of PAF in rat skin. PAF is a potent vasodilator that exerts its action in the cornea through G-protein modulated pathways. In rat skin, PAF induced plasma protein extravasation, increased transcapillary fluid flux and lowered Pif. In a rabbit DLK model, Holzer et al.[89] were able to demonstrate that the PAF receptor antagonist LAU-0901 measurably decreased both the incidence and severity of experimentally induced DLK. Bazen[42] interpreted this study to suggest that PAF released by corneal epithelium was a probable cause of DLK and that, by blocking the activity of PAF

through the use of a receptor blocker at the beginning of the cascade, DLK could be prevented.

Factors that Modulate Interstitial Fluid Pressure

An understanding of the factors that modulate interstitial fluid pressure has tremendous implications for detecting factors that might be useful in the prevention of trauma-induced tissue edema and its therapy. To that end, various substances have been demonstrated to either enhance the collagen contraction rate or limit the lowering of Pif.

> Understanding of the factors that modulate interstitial fluid pressure has tremendous implications for detecting drugs that might be useful in the prevention of DLK. To that end, various substances including α-Trinositol, platelet derived growth factor, endothelin have been demonstrated to enhance the collagen contraction rate induced by fibroblast activity and limit the lowering of Pif

As noted previously, platelet-derived growth factor (PDGF) and endothelin have been demonstrated to enhance the contraction process. Conversely, interleukin-1 and tumor necrosis factor-α have been shown to slow the rate of contraction while increased levels of intracellular cAMP will inhibit the contraction process.

The drug α-Trinositol, an anti-inflammatory agent that apparently works by altering intracellular calcium channels[146], has been demonstrated to either abolish or strongly attenuate the fall in Pif and edema formation in burns[147], dextran anaphylaxis[148], blockade of β1–integrins[149] and local frostbite injury[150]. It has also been demonstrated to have similar effects in the trachea in experimental asthma[151] and neurogenic inflammation[152].

Another potent stimulator of collagen gel contraction in vitro and in vivo is the BB isoform of platelet-derived growth factor (PDGF-BB). Rodt et al.[153] demonstrated that after dextran anaphylaxis had induced lowering of Pif in the rat paw, PDGF-BB injected subdermally brought the Pif completely back to normal. Linden et al. [154] found that the ability of PDGF-BB to counteract the tendency towards edema was accomplished by the stimulation of αVβ3-integrins.

The specific action of PDGF-BB have been further studied by Heuchel et al.[155] in mutant mice, wherein their group was able to demonstrate that the PDGF β-receptor operates via the phophatidylinositol 3'-kinase pathway and has a role in control of tissue fluid homeostatis in vivo by a direct effect on interstitial fluid pressure. Moreover, Ahlen et al.[146] demonstrated that cell-collagen interaction both in vivo and in vitro depend on phosphatidylinositol 3'-kinase, and that this dependence can be bypassed by a drug eliciting intracellular Ca^{2+} mobilization.

Inflammatory cytokines lower Pif and contribute to trauma-induced edema formation. This process is regulated by β-integrin mediated interactions between connective tissue cells and the extracellular matrix and is an active rather than a passive process. Change in Pif due to inflammatory cytokines involves the creation of cell-ECM tension translated from actin filaments found in the cell's cytoskeleton through β-integrin transmembrane proteins located in the cell membrane to collagen, fibronectin and laminin embedded in the proteoglycan matrix and is a G protein modulated process

The observation that both α-Trinositol and PDGF-BB demonstrate the ability to reverse the lowering of Pif, clearly suggest that the interstitial matrix is in a dynamic state wherein the tension on the connective tissue fibers can be increased and decreased within a few minutes to modify Pif and control tissue edema.

Summary of the Role of the Interstitium in Controlling Pif

In summary, acute inflammation and the production of inflammatory cytokines are associated with lowering of Pif and contribute to trauma-induced edema formation. This process is regulated by β-integrin mediated interactions between connective tissue cells and the extracellular matrix. This is an active rather than a passive process involving the creation of cell-ECM tension translated from actin filaments found in the cell's cytoskeleton through β-integrin transmembrane proteins located in the cell membrane to collagen, fibronectin and laminin embedded in the proteoglycan matrix.

> Understanding the manner in which Pif is controlled in tissue other than the cornea underscores the importance of cell-ECM tension. Application of this science to the cornea appears challenging because historically, the role of keratocytes has been primarily assigned to the synthesis of the extracellular matrix. A specific role for the keratocyte in the control of Pif or the production of cell-ECM tension has largely been ignored by corneal physiologists

Inflammatory cytokines such as prostaglandins like PGE_1, interleukin-1, PAF, LPS and TNF-α inhibit cell matrix contraction forces and induce tissue edema by allowing the ECM comprised of highly negative glycosaminoglycans to expand and drive Pif negative. Inflammatory cytokines interact with the cell by binding to G-protein and other receptors and affect intracellular levels of cAMP. Altering intracellular cAMP affects actin polymerization as well as other complex factors that influence cell shape, motility and function. Neurogenic stimulation and other chemokines produce similar effects. This effect is reversible. As the inflammation subsides, the effect of inflammatory cytokines is reduced. In the absence of inflammatory cytokines,

cell mediated contraction reverses ECM expansion and Pif is normalized.

Cell mediated ECM expansion can also be modulated and reversed by chemokine agents that interact with inositol processing, such as α-Trinositol or PDGF-BB. We also postulate that outside-in transmembrane signaling will affect cell-ECM tension and thereby affect Pif by creating conformational changes in β-integrin structure. Changes in β-integrin activity can in turn trigger the activation of cytoplasmic protein kinases, such as focal adhesion kinase (FAK) or Src. These kinases can then phosphorylate other proteins initiating a chain reaction that likely affects the cells cytoskeleton. Such outside-in signals transmitted by integrins can influence many aspects of cell behavior including tension, motility, growth and apoptosis.

Role of Cell-ECM Tension

An understanding of how interstitial pressures are controlled in tissue other than the cornea underscores the importance of cell-ECM tension. Application of this science to the cornea appears challenging because historically, the role of keratocytes has been primarily assigned to the synthesis of the extracellular matrix, collagen and other highly complex filamentary elements. A specific role for the keratocyte in the control of Pif or the production of cell-ECM tension has largely been ignored by corneal physiologists. In contrast, there is broad appreciation of the importance of highly specific configurations and arrangements of the collagen matrix as well as regional distribution of glycosaminoglycans to corneal form and function[156-158]. However, to attain such highly localized anatomical diversity, it is likely that keratocytes respond to local forces or chemokines that induce or regulate specific biosynthetic and behavioral pathways. The forces that cause keratocytes to produce such specific microenvironments or even assume their classic anisotropic orientation in the cornea are not well understood[159].

> Despite an apparent lack of investigation of the role of keratocyte-ECM tension on Pif, there is a significant body of work demonstrating that keratocyte-ECM tension plays an important role in wound healing. Studies indicate that corneal fibroblasts exert significant mechanical forces on the ECM

Despite an apparent lack of investigation of the role of keratocyte-ECM tension on Pif, there is a significant body of work demonstrating that keratocyte-ECM tension plays an important role in wound healing[159-162]. Studies on the role of keratocytes in corneal healing indicate that corneal fibroblasts do exert mechanical forces on the ECM. These forces are hypothesized to create a constant level of tension on the matrix of the cornea. Moreover, corneal fibroblasts cultured on a 3-dimensional collagen matrix respond to mechanical stress by increasing or decreasing tensional forces on the interstitium in order to maintain this state of tensional homeostasis. This response to mechanical compression or expansion of the surrounding matrix is rapid and directional. Cell movement,

ligand binding and the anisotropic orientation by fibroblasts suggest a significant role for integrins and the cell actin cytoskeleton in this process.

> Corneal fibroblasts cultured on a 3-dimensional collagen matrix respond to mechanical stress by increasing or decreasing tensional forces on the matrix in order to maintain a state of tensional homeostasis. This response is rapid and directional. Cell movement, ligand binding and the anisotropic orientation by fibroblasts suggest a significant role for integrins and the cell actin cytoskeleton in this process

Rapid corneal fibroblast response to matrix compression or expansion also suggests that fibroblast behavior is modulated by "outside-in transmembrane signaling" wherein external changes to the cell environment elicit immediate changes in fibroblast function. This cell behavior, likely induced via outside-in transmembrane signaling, has the effect of controlling the degree of mechanical force exerted on the ECM by the corneal fibroblast. Also, cell death and various substances such as cytochalasin D that disrupts the actinomyosin contractile apparatus, cause relaxation of transient cell-induced tension produced by cultured corneal fibroblast on the 3-dimensional collagen matrix[160,161]. In this manner, corneal fibroblasts demonstrate the capacity to create a constant tension on the ECM and that such tension can be rapidly modified by local cytokines and environmental stress. In addition, this tension appears to be a product of actin cytoskeleton tension communicated to the ECM by transmembrane proteins such as integrins.

> Corneal fibroblasts demonstrate the capacity to create a constant tension on the ECM and that such tension can be rapidly modified by local cytokines and environmental stress. In addition, this tension appears to be a product of actin cytoskeleton tension communicated to the ECM by transmembrane proteins such as integrins

In addition to an appreciation of the significant contribution of keratocyte-ECM tension to tensional homeostasis, there is important evidence that assigns a role for keratocytes in the control of tissue edema in the cornea. Corneal swelling in the presence of epithelial trauma has been demonstrated to occur in highly localized manner creating a preferential anterior swelling and a posterior thinning[163,164]. This finding is remarkable when one considers that it has been proposed that the cornea should tend to swell in the opposite direction based upon the tightly packed physical structure of local collagen networks anteriorly, the non-homogenous anterior to posterior distribution of GAG's in the ECM and the variation in glucose concentration between the anterior and posterior stroma [156,158]. The observation that epithelial trauma and release of inflammatory cytokines counteracts these latter powerful mechanisms suggests that the corneal stroma exerts additional direct control over Pif and that states of inflammation and trauma can disrupt factors regulating stromal hydration.

Ruberti et al.[163] and Karon and Klyce[164] attributed this effect to the release of osmotically active macromolecules into the ECM caused by keratocyte apoptosis induced by inflammatory cytokines.

> Corneal swelling in the presence of epithelial trauma has been demonstrated to occur in highly localized manner creating a preferential anterior swelling and a posterior thinning. These studies assign a significant role for keratocytes in the control of tissue edema in the cornea.

In summary, keratocytes and corneal fibroblasts appear to demonstrate behaviors that would allow them to participate in the control of Pif in the stroma. Corneal fibroblasts appear able to create specific levels of cell-ECM tension directed towards producing a state of corneal homeostatic tension. This force appears to involve the cell actin cytoskeleton, transmembrane proteins, ligand binding and outside-in transmembrane signaling. Localized corneal injury and associated inflammatory cytokines demonstrate marked effects on local tissue edema. It is postulated that inflammatory cytokines released into the corneal stroma create this effect by causing release of osmotically active macromolecules by keratocytes into the ECM.

Distribution of Interstitial Fluid Pressures in the Cornea

Universally recognized properties of corneal tissue include its unique transparency and the relationship between that characteristic and the need to maintain negative interstitial fluid pressures, compact collagen periodicity, tight junctions and constant ion pumping[156-158,165]. In essence, the core characteristic that makes the cornea transparent is a highly negative Pif. This factor significantly contributes to optical clarity and the relatively dehydrated state of corneal tissue. A negative Pif is also essential for the success of LASIK surgery as this is the principle force holding the flap in position following the procedure.

> The concept that interstitial fluid pressure or Pif in the cornea is not distributed in a homogenous fashion does not appear to be widely recognized in the field of refractive surgery or in the ophthalmologic literature

In our experience, the concept that interstitial fluid pressure or Pif in the cornea is not distributed in a homogenous fashion does not appear to be widely recognized in the field of refractive surgery or in the ophthalmologic literature. However, Wiig[166] using sensitive micropipette measuring techniques demonstrated that the interstitial pressure in the cornea in both human and animal eyes is highly negative compared to peripheral cornea, limbus and dermis. Wiig also demonstrated that corneal Pif is highly negative with the center of the cornea being nearly 40 to 50X more negative than dermis. Most importantly, the distribution of Pif is not uniform within the cornea itself with the central cornea being 30 to 40 X more negative than the limbus and at least 3-5X more negative than peripherally in the

> Corneal Pif is highly negative with the center of the cornea being nearly 40 to 50X more negative than dermis. Most importantly, the distribution of Pif is not uniform within the cornea itself with the central cornea being 30 to 40 X more negative than the limbus and at least 3-5X more negative than peripherally in the cornea (see Figure 1)

cornea (Figure 1). The localization of the most negative Pif at the corneal center likely provides the visual axis with the most compact collagen / proteoglycan / glycosaminoglycan (GAG) matrix and the highest level of optical clarity. Wiig[167] was also able to demonstrate that this marked gradient in interstitial fluid pressure from limbus to corneal center contributed to a directionality in bulk fluid flow from periphery to corneal center that was estimated to occur at approximately 120 nl/hr in the normal undisturbed rabbit cornea.

We believe that this Pif gradient and the mechanisms that influence fluid flux in the cornea are the primary drivers in a multitude of clinical effects surrounding

DLK, Stage 4 DLK, CTK, CFN, FNS and CLK are all caused by a loss of control of those factors regulating interstitial fluid pressures (Pif) and the Pif gradient in the cornea. Inflammation, inflammatory mediators, toxins and cell injury play a role in the etiology of these disorders but do so via their effects on the modulation of Pif. Alteration in Pif creates a cascade of physiologic processes that reorder corneal fluid dynamics, alter tissue compliance, redistribute tissue tension across the cornea and initiate reversible biologically mediated mechanical processes

excimer laser refractive surgery including DLK, Stage 4 DLK, CTK, CFN, FNS and CLK. In fact, we posit that DLK, Stage 4 DLK, CTK, CFN, FNS and CLK are all caused by a loss of control of those factors regulating interstitial fluid pressures (Pif) in the cornea. Inflammation, inflammatory mediators, toxins and cell injury play a role in the etiology of these disorders but do so via their effects on the modulation of Pif. Alteration in Pif creates a cascade of physiologic processes that reorder corneal fluid dynamics, alter tissue compliance, redistribute tissue tension across the cornea and initiate reversible biologically mediated mechanical processes. However, in order to develop such a disease model, we believe it instructive to explore these physiologic elements in greater detail.

A mechanism for creation of either the Pif gradient or the directionality of bulk fluid

flow is not understood. In the traditional view of control of corneal fluidics where everything related to control of Pif occurs as a result of endothelial activity, potential mechanisms might include a preferential effect of the endothelial pump centrally in the cornea compared to the periphery or a variation in proteoglycan or ECM composition across the cornea.

A mechanism for creation of either the Pif gradient or the directionality of bulk fluid flow in the cornea has not been understood.

The traditional view of control of corneal fluidics, where everything related to control of Pif occurs as a result of endothelial activity, is wholly inadequate and completely unable to explain this process.

Perhaps, in the field of refractive surgery there is no more important aspect of corneal physiology that requires understanding than the mechanism controlling the Pif gradient.

A variation in the effect of endothelial pump function may occur if: 1) central endothelial cells were more efficient than peripheral cells; 2) endothelial cell density varied across the cornea so that more cells were present centrally or; 3) if the effect of the endothelial pump on local tissue varied as a function of the distance of that tissue from the endothelial layer.

In corneal endothelium, pump efficiency and the metabolic function of individual

We suggest that there are five factors that create the Pif gradient and create directional bulk fluid flow.

1. Fluid inflow from the limbus
2. A compensatory increase in endothelial cell density peripherally such that the ion pump capacity counteracts this fluid influx
3. A compensatory increase in keratocyte population density peripherally such that the increased cell-ECM tension counteracts the tendency of the stroma to swell
4. Collagen interweaving to constrain the tendency of the peripheral cornea to swell due to fluid influx. This limit to tissue swelling slows the influx of fluid from the limbus
5. Progressive increase in the local effect of endothelial pumping on local tissue Pif from periphery to center as the stroma becomes increasing thinner. We propose that, in general, as distance from the endothelial cell layer decreases, local pump effect increases. Tissue closer to the endothelial cell layer would experience an increased local pump effect. This seems intuitive. The rate at which pump efficiency decreases with distance may be linear or non-linear.

cells are believed to be homogenous. However, there is limited scientific data to support that premise. Nevertheless, based on that assumption, variance in local pump efficiency from periphery to center could occur if there were marked differences in the density of localized cell populations. However, since the distribution in cell density of the endothelial layer has been shown to be lower centrally in the cornea compared to the periphery[168], it is unlikely that endothelial cell density is a cause for the Pif gradient unless one postulates that there is an inverse relationship between endothelial cell density and function. The understanding that loss of endothelial cells, in general, reduces pump effectiveness would not appear to support such an assumption.

The distribution of proteoglycans in the human cornea has been shown to vary from anterior to posterior[156-158]. However, there is no data to support the idea that there is a change in proteoglycan composition in tissue as one moves from the periphery of the cornea to the center. It would also be difficult to conceptualize a mechanism for how such a distribution might either create or maintain a Pif gradient. As a result, we believe variation in proteoglycan distribution from center to periphery to be an unlikely source of the Pif gradient in the cornea.

In the description of our complete model of control of corneal fluidics, we suggest that there are five factors that create this Pif gradient.

1. Fluid inflow from the limbus
2. A compensatory increase in endothelial cell density peripherally

such that the ion pump capacity counteracts this fluid influx

3. A compensatory increase in keratocyte population density peripherally such that the increased cell-ECM tension counteracts the tendency of the stroma to swell

4. Collagen interweaving to constrain the tendency of the peripheral cornea to swell due to fluid influx. This limit to tissue swelling slows the influx of fluid from the limbus

5. Progressive increase in the local effect of endothelial pumping on local tissue Pif from periphery to center as the stroma becomes increasing thinner. We propose that, in general, as distance from the endothelial cell layer decreases, local pump effect increases. Tissue closer to the endothelial cell layer would experience an increased local pump effect. This seems intuitive. The rate at which pump efficiency decreases with distance may be linear or non-linear.

Much like a military operation, the cornea places maximal effort on managing fluid influx at the limbus, which represents the primary point of entry for fluid in the undisturbed cornea. This allocation of maximal resources to the region serves to dehydrate the peripheral cornea as quickly as possible. As a consequence, corneal tissue can experience a progressive degree of dehydration from periphery to center. In addition, fewer endothelial cells and keratocytes are needed centrally to maintain a highly a negative Pif. The combination of these mechanisms ultimately produce the Pif gradient. In addition, a substrate that has the highest level of collagen compaction and fewest interstitial cells in the corneal center ultimately maximizes the clarity of central corneal optics.

Under normal homeostatic conditions we propose that the Pif gradient may be attributed to limbal fluid influx, peripheral collagen interweaving that limits tissue expansion that serves to slow fluid movement from the limbus, gradients in endothelial and keratocyte cell populations from periphery to center and localized endothelial pump effects

In support of this model for creation of the Pif gradient, keratocyte population density has been shown to vary from center to periphery and from anterior to posterior. However, similar to the endothelium, peripheral cornea exhibits increased keratocyte population density compared to the center[169]. Patel et al.[170] and Hahnel et al.[171] both demonstrated significantly fewer keratocytes in the posterior compared to the anterior stroma, while Hahnel et al.[171] showed that the volume density of cells in the posterior stroma was significantly larger than in the anterior and central stroma. In the context of the role of interstitial cells in controlling Pif elsewhere in the body, we suggest that these variations in population density and cell volume could assist in creation of centralized fluid flow directionality and a Pif distribution with the greatest degree of negativity located centrally.

Therefore, under normal homeostatic conditions we propose that the Pif gradient may be attributed to limbal fluid influx, peripheral collagen interweaving that limits tissue expansion that serves to slow fluid movement from the limbus, gradients in endothelial and keratocyte cell populations from periphery to center and localized endothelial pump effects. Of additional concern however is how the Pif gradient present under normal physiologic conditions is maintained in a system perturbed by refractive surgery.

> In a cornea with compromised epithelial barrier function and an expanding ECM, the rate that the endothelial pump can clear imbibed fluid from a local tissue will likely vary based upon proximity of the tissue to the endothelial layer, water binding properties of the tissue, Pif of the tissue and most importantly, changes in tissue compliance.

As it relates to DLK and related entities, it is equally important to consider the potential impact of proximity of tissue to the endothelial pump on the cornea's ability to maintain hydration at homeostatic levels when the cornea is subjected to surgical stress. In addition to distance between the pump and its target tissue, the instantaneous local effect of the endothelial pump will also vary as a function of the physical status of the cornea. Local variations in Pif, tissue compliance and tissue tension can significantly alter the ability of the endothelial pump to maintain or alter the hydration state of a specific tissue or localized area of tissue.

Under steady state conditions in a closed system with intact barrier function and ion pump physiology, the interstitial pressure moving from anterior in the cornea to posterior would likely be constant, irrespective of variable pump efficiency based upon tissue proximity. This is confirmed by the findings of Wiig[166]. However, in a system not in steady state with compromised epithelial barrier function and an expanding ECM, the rate that the endothelial pump can clear imbibed fluid from a local tissue will likely vary based upon:

1. Proximity of the tissue to the endothelial layer. Fluid closest to the endothelial cell layer will be removed before fluid located further away.
2. Physical properties including water-binding affinity of proteoglycans in the cornea. Dermatan sulphate, the predominant glycosaminoglycan distributed anteriorly in the human cornea demonstrates a different affinity for water than the Keratan sulphate located more posteriorly[156-158].
3. Interstitial fluid pressures of the local tissue. For example, tissue with a highly negative Pif will potentially pull fluid away from the endothelial pump, decreasing its efficiency
4. Barriers to directional fluid movement created by the LASIK interface. Lack of adhesive properties in the interface may allow fluid to accumulate in this potential space rather than move into adjacent tissue that is less compliant.
5. Changes in tissue compliance and volume created by altered tissue tension. In the post refractive surgery eye, tension on collagen fibrils is redistributed. Increased tissue tension will reduce the volume of a tissue and

decrease tissue compliance. Conversely, decreased tension will increase tissue compliance and tissue volume.

These factors and other changes in pump metabolism could adversely affect the rate at which the state of hydration of a local area of the cornea can change in the face of a disruption or restoration of barrier function.

Detailed Description of the Model Components

In light of this information, we propose a model for DLK, Stage 4 DLK, CTK, CFN, FNS and CLK that postulates that inflammatory cytokines, pharmacologic agents and injury to keratocytes disrupt stromal regulated homeostasis of Pif by disrupting ECM-keratocyte interactions. The disruption of these interactions leads to matrix relaxation and drives Pif more negative. The distribution of Pif can be highly localized in the cornea. The force generated draws fluid and inflammatory cells into the interface and stroma and leads to the formation of tissue edema. Alteration in Pif further creates a cascade of physiologic processes that reorder

corneal fluid dynamics, alter tissue compliance, redistribute tissue tension across the cornea and initiate reversible biologically mediated mechanical processes (Figures 3 through 5).

Corneal refractive surgery results in the synthesis and release of numerous inflammatory mediators. These inflammatory cytokines are produced as a result of many factors including tissue injury, toxic chemicals, actions of pharmacologic agents, thermal damage and the introduction of exogenous antigens

DLK, Stage 4 DLK, CTK, CFN, FNS and CLK are caused by the effects of inflammatory cytokines, pharmacologic agents and injury to keratocytes. These factors disrupt stromally regulated homeostasis of Pif by disrupting ECM-keratocyte interactions leading to matrix relaxation and a more negative Pif. The force generated draws fluid and inflammatory cells into the interface and leads to the formation of tissue edema. Alteration in Pif creates a cascade of physiologic processes that reorder corneal fluid dynamics, alter tissue compliance, redistribute tissue tension across the cornea and initiate reversible biologically mediated mechanical processes (see Figures 3 through 5)

Tissue Trauma and the Release of Inflammatory Mediators

Corneal refractive surgery results in the synthesis and release of numerous inflammatory mediators[172]. These inflammatory cytokines are produced as a result of many factors including tissue injury, toxic chemicals, actions of pharmacologic agents, thermal damage and the introduction of exogenous antigens. For example, laser refractive surgery such as PRK, LASIK or LASEK involves acute injury to the cornea, conjunctiva and sclera and is frequently combined with the exogenous introduction of inflammatory chemokines. In the process, epithelial and interstitial cells in the cornea and conjunctiva are injured causing cell destruction and apoptosis. These factors lead to the presence of significant, albeit variable, amounts of inflammatory mediators and chemokines in the ocular milieu.

A mechanical microkeratome provides several avenues leading to the introduction or release of inflammatory chemokines. Lipopolysaccharide biofilms and sterile immunogenic debris can be harbored on the device and transferred directly to the conjunctiva and cornea. Physical destruction of conjunctival epithelium due to vacuum forces and crush injury to the conjunctiva, vasculature and sclera also occur. The oscillating blade cuts through corneal epithelium and stroma as well as applying shearing trauma to the epithelial layer. Varying sizes of epithelial defects are common.

Infrared lasers operating at the femtosecond range have recently been utilized in LASIK surgery. The use of sterile single use vacuum rings, lower vacuum levels and the elimination of epithelial tissue shear likely reduce tissue injury and the deposition of inflammatory material. However, these devices dissect corneal tissue through the use of short duration high-energy laser pulses that cause tissue photodisruption[173]. The intensity of the energy delivered to the epithelium and stromal tissue during light induced optical breakdown converts solid cell structure to plasma. Photodisruption and the creation of plasma produce a local thermal load and release large acoustic and sonic shockwaves into the adjacent tissue. We propose that these shockwaves and thermal load delivered to the keratocyte-ECM complex cause tissue injury, induce inflammatory cytokines and create outside-in transmembrane signaling.

> Shockwaves and thermal load delivered to the keratocyte-ECM complex by femtosecond laser pulses cause tissue injury, induce inflammatory cytokines and create outside-in transmembrane signaling. This is the primary mechanism causing DLK in the use of femtosecond laser technology

Experience with femtosecond lasers indicates that, in general, reducing the amount of energy released in the corneal tissue markedly reduces side effects from the treatment. Specifically, the incidence and severity of DLK is decreased with reduction and optimization of laser energy[43,79,80,100] and refractive accuracy in the early postoperative period is markedly better (Will BR. Diffuse Lamellar Keratitis and the IntraLASIK Procedure. Intralase™ Users Meeting, Dana Point, California, June 2002). Hu et al.[174] have shown decreasing amounts of keratocyte activation and density of interface particles with reduced femtosecond laser energies. In rabbit eyes, Kim et al.[175] found a more pronounced inflammatory response after femtosecond flap creation compared to a microkeratome. Will demonstrated that optimizing laser spot separation and laser pulse energy to a finely tuned harmonic could similarly reduce adverse side effects by creating a continuous photodisruption plane that was proposed to reduce the injurious effect of the thermal load and sonic shockwave locally (Will BR. Continuous Plane

Photodisruption – Creating the Ideal Flap using Optimized Laser Parameters. Intralase™ Users Meeting. New Orleans, Louisiana. October 2004).

Role of Inflammatory Mediators

Inflammatory mediators such as IL-1, PAF and TNF-α have been suggested to play a role in the pathogenesis of DLK[20,21,42,54,80,89,172]. However, prior to the introduction of our model, the proposed role of such cytokines has been to attract PMN's into the LASIK interface where they can release matrix

> In our model, inflammatory chemokines released or synthesized as a consequence of laser refractive surgery do not initiate a cascade that leads to tissue necrosis. Rather, these chemical messengers play a role in affecting the control of interstitial fluid pressure in the cornea by binding to receptors on keratocyte membranes. Upon binding, G-protein mediated changes in the keratocyte cytoskeleton are initiated causing relaxation of beta-integrin ECM tension in the stroma. As the ECM is allowed to expand, interstitial fluid pressures become increasingly negative. Combined with disruption of tight junctions in the corneal epithelium, this causes the stroma to imbibe fluid

metalloproteinases and collagenases that initiate tissue destruction, cause upregulation of the production of

additional chemokines by stromal keratocytes as well as to induce keratocyte apoptosis. This inflammatory cascade has been proposed as a primary cause of DLK. Although we agree that this pathophysiologic cascade proposed by these and other authors may occur to a very limited degree, it is not the primary cause for the predominant pathologic findings in these disorders.

In our model, inflammatory chemokines released or synthesized as a consequence of laser refractive surgery do not initiate a cascade that leads to tissue necrosis. Rather, these chemical messengers play a role in affecting the control of interstitial fluid pressure in the cornea (Figure 4). This is accomplished by the binding of these chemokines to receptors on keratocyte membranes. Upon binding to appropriate receptors including G-protein coupled receptors, G-protein mediated changes in the keratocyte cytoskeleton are initiated causing relaxation of beta-integrin ECM tension in the stroma. As the ECM is allowed to expand, interstitial fluid pressures become increasingly negative. Combined with disruption of tight junctions in the corneal epithelium, this causes the stroma to imbibe fluid.

Under normal circumstances in the postoperative period, this effect is clinically insignificant due to the compliance characteristics of the interstitial space, the capacity of the endothelial pump to keep pace with fluid uptake and the rapid restoration of epithelial barrier function. Mild sub-clinical edema may induce minor effects on refractive errors producing transient refractive overshoots during the first few weeks of healing. However, as control over Pif homeostasis is attained, these effects are typically unimportant to the surgical result.

However, in cases where significant tissue injury has occurred (such as a large epithelial defect), the surgical trauma is excessive or a bolus of exogenous inflammatory chemokines are introduced intraoperatively, the acute effects on control of interstitial fluid pressures in the cornea can overwhelm compensatory mechanisms. Inflammatory chemokines may recruit acute inflammatory cells to the tear film. Fluid uptake by the corneal stroma causes fluid from the tear film to be drawn into the interface as tight junctions have been disrupted. Depending upon the negativity of Pif and the cell load in the tear film, inflammatory cells may appear in the interface. If the cellular response is large the clinical appearance will be designated to be DLK. Conversely, if there are only a few cells present, the clinical designation will be CTK. In either case, the pathophysiologic cascade in the stroma is identical (Figure 3).

> In DLK, the mechanism causing the pathognomonic shifting sands appearance is the bulk movement of fluid towards the corneal center rather than a chemotactic effect of antigens deposited in the interface. This process creates the pathognomonic "Sands of the Sahara" appearance in the LASIK interface

In the DLK phenotypic presentation, the mechanism causing the pathognomonic shifting sands appearance is the bulk movement of fluid towards the corneal center rather than a chemotactic effect of antigens deposited in the interface. Because there is a gradient in interstitial fluid pressures from periphery to center, it is likely that fluid moves centrally until it is imbibed by adjacent stroma. Once the compliance of the ECM space is saturated locally, fluid no longer moves into adjacent stroma but rather is moved closer to the corneal center. This, in combination with lid compression during blinking, creates waves of fluid movement that sorts any inflammatory cells in the interface moving towards the center into "windrows". This process in turn creates the pathognomonic "Sands of the Sahara" appearance in the LASIK interface.

> The corneal center where Pif is highly negative swells. As the compliance of the interstitial space of the flap becomes saturated, frank tissue edema occurs causing increased separation of collagen fibrils, loss of collagen periodicity and the opacification of the stroma. If severe, macrofolds in the LASIK flap will occur causing severe loss of BCVA

The corneal center where Pif is highly negative swells as fluid moves along the LASIK interface. As the compliance of the interstitial space of the flap becomes saturated, frank tissue edema occurs. Clinically, this is demonstrated by increased separation of collagen fibrils, loss of collagen periodicity and the opacification of the stroma. If severe, frank folds in the LASIK flap will occur causing clinically evident full thickness macrofolds. Both events are accompanied by severe loss of BCVA. Once the compliance of this anterior central stromal compartment is saturated and epithelial tight junctions have recovered, fluid movement into the stroma is decreased. However, the edema and chemokine load results in marked keratocyte apoptosis in the anterior stroma and LASIK flap. As a result, despite a fully functioning endothelial pump, without an interstitial cell population capable of creating functional cell-ECM tension, the frank edema in the anterior stroma and flap does not rapidly resolve. As a consequence, the corneal opacification and loss of BSCVA can last for months.

Role of Excimer and Infrared Femtosecond Lasers

The contribution of excimer and infrared Femtosecond lasers to the incidence of DLK and CTK has only been rarely considered. However, there is significant evidence to support a role for these devices in the causation of these disorders.

The effect of excimer lasers in the etiology of these entities is frequently ignored. However, if we examine the literature regarding interface inflammation in other corneal lamellar procedures there is a near complete absence of reports of DLK or CTK in corneal lamellar transplants and in myopic keratomileusis insitu (MKM). It is reasonable to consider whether laser induced thermal

energy may play a role in causing DLK as this element is absent in these other lamellar procedures. Gunn et al. (JL Gunn, SL Forstot, A Hatsis et al. "Sands of the Sahara: Post LASIK Interface Inflammation – Reality of Mirage?" Poster: Symposium on Cataract, IOL and Refractive Surgery, San Diego, CA, USA, April 1998) observed that the density of inflammatory cells in the interface varied as a function of amount of excimer laser pulses delivered. Certainly, in the context of currently proposed pathophysiologic mechanisms involving antigens and toxins, it is difficult, if not impossible, to explain why other corneal lamellar procedures have not been afflicted by DLK.

Both excimer and femtosecond lasers contribute to the formation of DLK and related entities. Keratocyte apoptosis in high raster energy femtosecond laser corneal surgery and within the footprint of the excimer laser is directly related to outside-in transmembrane signaling caused by the increased thermal loading and sonic shockwaves. Keratocyte injury and apoptosis adversely affects the control of Pif and contributes significantly to the development of DLK, Stage 4 DLK, CTK, CFN, FNS and CLK

Use of infrared femtosecond lasers has been linked to DLK[43,79-83]. All of the cases in the series reported here involved use of infrared femtosecond lasers. It has also been our observation that reducing raster and sidewall energy is associated with marked decrease in both the incidence and severity of DLK. Will suggested that DLK was associated with keratocyte injury and loss of control of Pif in the corneal stroma (Will BR. Management of Stage 4 Diffuse Lamellar Keratitis. Scientific Session. American Society of Cataract and Refractive Surgery. Washington DC. April, 2005).

Netto et al.[176] demonstrated that the amount of keratocyte apoptosis is higher with LASIK flap creation using an 15 kHz infrared Femtosecond laser compared to traditional microkeratome. In addition, there is significantly less keratocyte apoptosis with the 30 and 60 kHz laser compared to the 15 kHz. Typically, raster pulse energy and spot separation is much less with the 30 and 60 kHz than the 15 kHz system which likely accounts for this finding. Similarly, Hu et al.[174] found a decrease in the number of activated keratocytes in the stroma when lower laser pulse raster energies were used.

We believe that keratocyte apoptosis, and particularly keratocyte apoptosis in high raster energy femtosecond laser corneal surgery or the footprint of a broad beam excimer laser, is directly related to outside-in transmembrane signaling caused by the increased thermal loading and sonic shockwaves. We suggest that this keratocyte injury and apoptosis adversely affects the control of Pif and contributes significantly to the development of DLK, Stage 4 DLK, CTK, CFN, FNS and CLK.

Importance of Outside-in Transmembrane Signaling

Outside-in transmembrane signaling may interfere with cell-ECM adhesion. Petroll et al.[160], Vishwanath et al.[161] and Roy et al.[162] clearly demonstrated the impact of outside-in transmembrane signaling on the ability of corneal fibroblasts to exert pressure or tension on the ECM. We posit that the effect of excimer laser and infrared femtosecond lasers is to cause alterations in the control of Pif in the cornea by interfering with outside-in transmembrane signaling in the corneal stroma.

> Outside-in transmembrane signaling is likely the mechanism by which lasers contribute to the creation of both DLK and CTK, however outside-in transmembrane signaling is likely the primary mechanism behind the development of CTK

Outside-in messaging has been demonstrated to occur in immune cell adhesion wherein shear forces on β-integrins at the piconewton range can initiate changes in cell morphology[177-179]. Similarly, corneal tissue compression and relaxation caused by excimer laser pulses, thermal tissue loading and the sonic shockwave from infrared femtosecond lasers may disrupt binding between collagen and fibronectin ligands in the cornea and β-integrins on keratocytes. Disruption or uncoupling of these ligand--β-integrin bonds may cause the β unit of the integrin complex to revert to its inactive and folded state. In the process of converting from an active to inactive conformation, the β-integrin may cause initiation a FAK or Src cascade within the keratocyte. Such a cascade could in turn affect the cell cytoskeleton. Changes in the cytoskeleton may interfere with cell-ECM or cell-cell interaction or induce keratocyte apoptosis in the cornea leading to expansion of the ECM and increased negativity of the Pif.

> The primary mechanism causing CTK is outside-in transmembrane signaling mechanism and is not caused by inflammatory chemokines and G protein processes. As a result, CTK does not respond to topical steroid or non-steroidal anti-inflammatory medications (NSAID's). In CTK there is limited recruitment of inflammatory cells to the tear film and interface since production of inflammatory cytokines is minimal. Outside-in transmembrane signaling caused by excimer and femtosecond lasers creates a majority of the pathology in the center of the cornea, which is where the preponderance of pathology occurs in DLK and CTK

Outside-in transmembrane signaling is likely the mechanism by which lasers contribute to the creation of both DLK and CTK, however outside-in transmembrane signaling is likely the primary mechanism behind the development of CTK (Figure 5). Because the outside-in transmembrane signaling mechanism is not caused by inflammatory chemokines and G protein processes,

CTK does not respond to topical steroid or non-steroidal anti-inflammatory medications (NSAID's). There is also limited recruitment of inflammatory cells to the tear film and interface since production of inflammatory cytokines is minimal. Outside-in transmembrane signaling caused by excimer and femtosecond lasers would also be expected to create a majority of the pathology in the center of the cornea, which is where the preponderance of pathology occurs in DLK and CTK. As a consequence, it is not necessary to postulate that a 'photoactivated' toxin is a factor in the etiology of CTK as such a molecule would be superfluous to the causation of this disorder.

Physiological Mechanisms creating Late Onset DLK

In the context of current disease paradigms involving antigens and tissue toxins, late onset DLK has been an enigma. It is difficult to explain how DLK can occur weeks or months after an uncomplicated LASIK surgery that exhibited no signs of interface inflammation during the initial postoperative period, without the acute introduction of either antigens or toxins.

Our model posits that sporadic and late onset cases of interface inflammation are caused by the same pathophysiologic mechanism that causes epidemic DLK and CTK. In our model, late onset and sporadic DLK or CTK occur primarily as a consequence of corneal epithelial defects and conjunctival trauma that in turn cause a release of inflammatory cytokines and mediators into the eye.

These cytokines diffuse into the stroma through breaks in the epithelial barrier and cause relaxation of cell-ECM tension through G protein mediated processes. As Pif becomes more negative in the stroma, fluid and inflammatory cells present in the tear film are absorbed through breaks in tight junctions within the epithelial barrier and accumulate in the old LASIK interface. Fluid influx may cause focal corneal edema, as has been observed in a number of cases of late onset DLK[180-182]. The accumulation of acute inflammatory cells in the interface in these late onset cases is likely due to the interface representing a path of least resistance as well as fibronectin deposited there[83,183] acting as a guide and "bread crumb trail" that facilitates inflammatory cell motility and migration. Fibronectin has as well known role as a guide element or scaffold that facilitates cell migration during embryogenesis[114].

Late onset cases of interface inflammation are caused by the same pathophysiologic mechanism that causes epidemic DLK and CTK and occur as a consequence of corneal epithelial defects and conjunctival trauma that cause a release of inflammatory cytokines. These cytokines diffuse into the stroma and cause relaxation of cell-ECM tension and lower Pif through G protein mediated processes. Fluid and inflammatory cells accumulate in the old LASIK interface and may cause focal edema. The accumulation of cells in the interface is due to the interface representing a path of least resistance and fibronectin acting as a guide and "bread crumb trail" that facilitates inflammatory cell motility and migration

The Hyperopic Shift

Defining Biomechanics

In order to fully appreciate the etiology and mechanism of the hyperopic shift in Stage 4 DLK and CTK as well as the reversibility of this phenomenon, an understanding of the interaction between biological and mechanical factors in the post LASIK cornea is required. Biomechanics of the cornea has been a topic of discussion and modeling for

> Biomechanics is not merely the study of the physical structure of biological systems. Labeling mechanical processes as "biomechanics" while ignoring fundamental biological process that occur conjointly in the tissue is simplistic and inaccurate.

many years. Roberts proposed a biomechanical model of the cornea in the post refractive surgery state and compared this model to clinical observations of Orbscan® II data[184-186]. Others have promoted similar concepts[187]. The primary clinical implication of the Roberts model is that it identified a definite trend in myopic LASIK to produce a higher degree of peripheral corneal steepening and a higher degree of central flattening than expected. In the Roberts model the principle forces considered to be acting in the tissue were predominantly mechanical in nature. Most importantly, the Roberts model provided no identifiable mechanism for reversibility.

We suggest that the study of biomechanics must not be limited to the mechanical effects of severing collagen filaments and the expansile forces of highly negatively charged proteoglycans under reduced states of tension or strain on the matrix produced by crosslinked fibrils. Labeling mechanical processes that are attributed solely to a material's physical structure and properties that happen to occur in biological systems as "biomechanics", while ignoring fundamental biological process that occur conjointly in the tissue, appears to us to be both simplistic and inaccurate. We believe this to be a faulty interpretation and application of the terminology and suggest that where the scope of a scientific discussion is limited to purely physical properties of the matrix, such a process be referred to as 'corneal mechanics'. Rather, it is our position that biomechanics studies the relationships between biological events or forces and the mechanical properties of physical structures. It is this broader based analysis of the interaction between biology and mechanics that ultimately defines and integrates tissue physiology with biophysical compliance. Perhaps the term 'biophysiologic mechanics' or the 'mechanics and physiology of biosystems' more accurately defines such a perspective.

As has been described in this paper, the contribution of biology in the post LASIK cornea is highly significant in the etiology of DLK, Stage 4 DLK, CTK, CFN, FNS

The contribution of biology in the post LASIK cornea is highly significant in the etiology of DLK, Stage 4 DLK, CTK, CFN, FNS and CLK and these biological forces create measurable mechanical tissue effects. Because the cornea experiences loss of control of Pif followed by gradual recovery of Pif homeostasis, the mechanical effects created by those biological forces also exhibit a pattern of reversibility with clinical worsening followed by recovery

and CLK. Our model also predicts that these biological forces create measurable mechanical tissue effects. Because the biology is dynamic, with the cornea successively experiencing loss of control of Pif followed by gradual recovery of Pif homeostasis, the mechanical effects created by those biological forces also exhibit a pattern of reversibility with clinical worsening followed by recovery. Although not described in any detail in this paper, our model also predicts that the biological impact of mechanical shape alteration in the cornea is also a significant factor contributing to the predictability, or lack thereof, of the refractive endpoint of laser vision correction surgery as well as much of the transient refractive errors found during the immediate postoperative period.

Of significant importance is the fact that the biomechanical events described in our model are derived by interpretation of actual Optical Coherence Tomography (OCT) and Pentacam™ tomography

investigations of cases of Stage 4 DLK, rather than on supposition and theory (Figure 9 through 11). Moreover, our understanding of tissue changes documented by these sophisticated imaging technologies, combined with the clinical findings of our patients, and our corneal model are derived by applying in a practical manner known physiologic principles and predicting how alterations in Pif and the resulting relationship between biological and mechanical forces in the cornea following LASIK will affect tissue behavior and shape.

Overview of Corneal Biomechanics

In our model, the hyperopic shift that occurs in Stage 4 DLK, CTK, CFN, FNS and CLK is due to interrelated biological and mechanical forces that ultimately

The hyperopic shift that occurs in Stage 4 DLK, CTK, CFN, FNS and CLK is due to interrelated biological and mechanical forces that ultimately affect corneal shape and refractive power. These biomechanical forces are directly and predominantly affected by characteristics of the "compliance" of the corneal extracellular compartment

affect corneal shape and refractive power. These biomechanical forces are directly and predominantly affected by characteristics of the "compliance" of the corneal extracellular compartment. Pif is a critical, but not the only, factor affecting tissue compliance. Rather, compliance of the corneal ECM tissue is determined by the relationship between tissue tension and Pif. Changes induced in the compliance of the compartment can be reversible or permanent depending upon the nature of the mechanisms affecting corneal tissue and these changes in compliance affect the shape and refractive power of the cornea. Induced changes in tissue compliance may also be asymmetric, in which case they may induce astigmatism and other higher order aberrations in the refraction.

> Pif is a critical, but not the only, factor affecting tissue compliance. Rather, compliance of the corneal ECM tissue is determined by the relationship between tissue tension and Pif. Changes induced in the compliance of the compartment affect the shape and refractive power of the cornea and can be reversible, permanent and asymmetric

Our model predicts that physiologic changes in the control of fluid dynamics in the cornea induced by laser refractive surgery that serve to increase hydration of the cornea postoperatively (such as inflammatory cytokines and outside-in transmembrane signaling) tend to result in hyperopic shifts and increased positive spherical aberration. In the converse, postoperative events that dehydrate the cornea tend to produce myopic shifts and likely reduce positive spherical aberration. Since it is generally much easier to create processes that increase local tissue hydration rather than inducing dehydration (since the cornea is already quite dehydrated under normal

environmental conditions), in the clinical setting we are most familiar with the former.

> Refractive effects result from localized changes in tissue compliance. Loss of control of tissue fluid dynamics creates changes in local tissue hydration.
>
> - Excessive local hydration results in central corneal flattening and peripheral corneal steepening and a hyperopic shift
> - Local tissue dehydration results in less corneal flattening and less peripheral steeping than normal producing a myopic shift

In our model, factors such as inflammation that disrupt Pif and cause a loss of control of tissue fluid dynamics create excessive local tissue hydration in a relatively predictable fashion. Excessive local hydration results in an unexpected increase in central corneal flattening and an unexpected increase in peripheral corneal steepening. If the forces that are influencing fluid dynamics are transient, these changes are reversible. It is these events that create a hyperopic shift, frequently combined with induced astigmatism and refractive overshoot. Similarly, processes in the postoperative period that cause excessive local tissue dehydration or do not result in the typical amount of tissue inflammation result in less corneal flattening and less peripheral steeping than normal. This produces a myopic shift and refractive overshoot. Our model predicts that these refractive effects occur directly as a result of localized changes in tissue compliance.

Control of the size and compliance of the corneal compartment

We posit that both before and after LASIK surgery, Pif and tissue tension represent the primary variables that control size and compliance of the corneal compartment. For purposes of this paper, we use the terms tissue tension synonymously with

> Pif and tissue tension represent the primary variables that control size and compliance of the corneal compartment. Tissue compliance (ΔVol / ΔPif) follows a linear relationship. Tissue tension determines the slope of this relationship by influencing the amount of change in tissue volume for a change in Pif

tissue strain or tissue loading. In general, compliance of a tissue follows a more or less linear relationship between change in tissue volume for a given change in Pif (ΔVol / ΔPif). In our model, tissue tension determines the slope of this relationship by influencing the amount of change in tissue volume for a change in Pif (Figure 6). This seems intuitive. However, this relationship between Pif and tissue tension has been confirmed by studies on the effects of uniaxial strain on dermis. In general, tissue under high degrees of tension will exhibit a low compliance while tissue under low degrees of strain will demonstrate high compliance[188].

In our model, in the undisturbed normal cornea, tissue tension throughout the cornea is relatively uniform. Therefore the relationship between ΔVol / ΔPif remains linear, predictable and generally consistent throughout the tissue. Within the physiologic range of Pif and tension in a normal cornea, the resulting interaction between ΔVol / ΔPif produces a medium suitable for clear vision that is capable of resisting the effects of external forces seeking to either hydrate or dehydrate the tissue or cause mechanical deformation of corneal shape. Collagen fibrils resist stretching while the ECM resists compression. Tissue hydration is predictable and is controlled by a feedback loop directed by the actions of endothelial cells, keratocytes and epithelium.

Forces that induce biomechanical change

In our model, corneal surgeries such as PRK, LASIK and LASEK alter Pif. However, in addition to transient local changes in Pif, laser refractive surgery also markedly affects;

1. The linearity of the ΔVol / ΔPif relationship
2. The rate of change of volume for a given change in Pif (slope of the compliance curve)
3. The distribution of tissue tension

Alterations in Pif, tissue tension and change in the slope of the compliance curve induce biological and mechanical changes in the cornea that directly cause predictable biomechanical events (Figure 6).

Loss of Linearity of the ∆Vol / ∆Pif relationship

In the normal pre-surgical or typical post-LASIK or post-PRK cornea, Pif is negative and the ∆Vol / ∆Pif relationship is linear. In the typical post-refractive surgery cornea this level of stability occurs because the control system for maintaining Pif is robust compared to the surgical insult. However, if the surgical event induces changes in fluid flux in the cornea that overwhelm normal homeostatic mechanisms, Pif and local fluid dynamics become pathologic. Mechanisms that maintain tissue clarity and tissue shape may become unstable and unbalanced if changes in tissue tension or Pif are significant in the post-surgery period.

> As Pif becomes positive and tissue tension is markedly reduced, the linearity of the ∆Vol / ∆Pif relationship is lost. This causes marked tissue edema

If Pif becomes positive and tissue tension is markedly reduced, as in a pathological condition such as Stage 4 DLK, CTK, CFN, FNS and CLK, the linearity of the ∆Vol / ∆Pif relationship can be lost. Under conditions of reduced tension, tissue exhibiting a positive Pif may take up significantly more fluid and exhibit marked swelling with no further change in Pif. Typically, when the latter occurs the tissue is visibly edematous. Under such pathologic conditions in grossly edematous tissue, normal mechanisms such as cell-ECM interactions that control hydration under physiologic conditions are absent. However, if the tissue can recover normal cell-ECM control processes, such edema may be reversible

> A positive Pif combined with loss of linearity of the ∆Vol / ∆Pif relationship is the primary event that occurs in the corneal flap thereby creating:
>
> • Central focal whitening
> • Central micro and macrofolds
> • Central flap thickening

over time. A positive Pif combined with loss of linearity of the ∆Vol / ∆Pif relationship is the primary event that occurs in the corneal flap during Stage 4 DLK, CTK, CFN, FNS and CLK and creates: 1) central focal whitening notable at the slit lamp; 2) central micro and macrofolds and; 3) central flap thickening. This same central whitening and striae effect can occur in the anterior stroma in PRK as has been noted by Sonmez and Maloney in their description of CTK[22].

Effects of Tissue Tension on Compliance

In our model, tension in the cornea markedly affects the compliance of the extracellular matrix (Figure 6). Specifically, as tension increases, the slope of the relationship between ∆Vol and ∆Pif becomes steeper and compliance

Tension affects the compliance of the ECM. Increasing tension reduces the amount of change in tissue volume for a change in Pif. If tension is very high, the tissue demonstrates little if any compliance Conversely, as tissue tension decreases, compliance increases and the amount of change in tissue volume for a given change in Pif increases. If tension is significantly reduced and the tissue exhibits minimal ECM cross-linking, the volume of the extracellular compartment may increase even if Pif remains constant

of the compartment decreases. Simply stated, increasing tension has the effect of reducing the amount of change in tissue volume for a specific change in Pif. If tension is very high the tissue effectively demonstrates little if any compliance and the volume or size of the extracellular compartment may actually decrease even if Pif remains unchanged. Conversely, as tissue tension decreases, compliance of the compartment may increase such that the amount of change in tissue volume for any given change in Pif increases. If tension is significantly reduced and the tissue exhibits minimal ECM cross-linking, the volume of the extracellular compartment may increase even if Pif remains constant.

LASIK and PRK alter the Pif, Tissue Volume and Tissue Tension Relationship

In our model, changes in Pif, tissue tension and tissue compliance create marked transient changes in corneal shape and refractive power. However, in order to appreciate how Pif and tension influence tissue compliance and how tissue compliance creates mechanical forces that affect corneal shape and refractive power characteristics, it is essential to understand how Pif and tension change from the preoperative to the postoperative condition in LASIK or PRK.

In addition to Pif and tension, fluid dynamics in the post-LASIK cornea are influenced by a number of important factors:

1. Loss of the epithelial barrier
2. Temporarily reduced capacity of the endothelial pump
3. Alterations in keratocyte behavior and keratocyte apoptosis that create local changes to Pif
4. Local increased negativity of Pif in the anterior cornea
5. Increasing rates of fluid influx due to local Pif negativity
6. Water binding by glycosaminoglycans

The complex interaction between these forces induce the reversible hyperopic shift

In the interest of simplicity, we will confine our comments to LASIK. However, the same principles clearly apply to PRK as well.

Specifically, LASIK surgery radically changes tissue tension and Pif from the pre-LASIK condition. Shifts in Pif and tension have the effect of changing the relationship between tension and Pif that is present in a state of homeostatic balance in the pre-LASIK cornea.

In addition to Pif and tension, fluid dynamics in the post-LASIK cornea are influenced by a number of important factors:

1. A breakdown in the epithelial barrier. Fluid uptake by the cornea is markedly increased thereby creating local edema. Local edema challenges the ability of corneal mechanisms to maintain a normal level of local tissue hydration.

2. A reduced capacity of the endothelial pump. Although endothelial cell loss is clinically insignificant, gross changes to endothelial cell morphology are well documented[189-200]. We believe that it is likely that total pump capacity and integrity of endothelial tight junction are temporarily compromised.

3. Inflammatory cytokines and chemokines cause both alterations in keratocyte behavior and keratocyte apoptosis. These changes occur most significantly near the corneal surface where the flap is created and the tissue is reshaped by an excimer laser. This creates local changes to Pif that further induce stromal edema.

4. If Pif becomes more negative in the anterior cornea as a result of the effects of inflammatory cytokines, this will interfere with the ability of the endothelial pump to remove excess fluid from that region, thereby counteracting its local influence to dehydrate the tissue.

5. Increasing rates of fluid influx occur anteriorly due to local Pif change. Fluid influx combined with a decremental loss or decay of local ion pump effect created by the distance a specific tissue is located away from the endothelial monolayer will result in fluid accumulation in the anterior cornea to a proportionately greater degree than normal

6. Glycosaminoglycan and proteoglycan distribution in the typical cornea exhibit a preponderance of dermatan sulfate anteriorly, which exhibits a greater affinity for binding water than the keratan sulfate found more posteriorly[156-158]. This further exacerbates the propensity for water hoarding in the anterior cornea.

It is the complex interaction between these forces and their combined effect on tissue compliance that induce the reversible hyperopic shift and other pathology observed in Stage 4 DLK, CTK, CFN, FNS and CLK.

In order to simply this relationship for purposes of discussing the model, we have divided the post-LASIK cornea down into four (4) distinct zones or tissue compartments (Figure 7). In reality, the transition between these zones is gradual. As further basic science research explores these mechanisms in greater detail, it may become necessary to add additional elements or even zones to the model. However, we believe that the four zones described herein encompass the primary factors that typically influence the relationship between Pif, tissue tension, tissue compliance and fluid dynamics in inflammatory disorders such as Stage 4 DLK, CTK, CFN, FNS and CLK and to a lesser extent during laser refractive surgery in general.

We will outline the key characteristics of each zone followed by an in-depth analysis of the mechanisms that occur in each zone. The four distinct response zones or compartments in the post-LASIK cornea and the specific changes in Pif, tension, compliance and fluid flux that occur in each compartment include (Figure 7 and Figure 8):

1. Zone 1 - The LASIK flap

   a. Tension in the entire flap tissue is markedly reduced
   b. Pif, which is already highly negative centrally, is driven more negative locally by inflammatory cytokines
   c. Due to the normal Pif gradient (wherein Pif is most negative in the corneal center), the local production of inflammatory cytokines by damaged epithelium and keratocytes and an expanding ECM, fluid will tend to move through the epithelial break located at the flap sidewall towards the center of Zone 1 along the LASIK interface

   d. The slope of the compliance curve is decreased due to markedly reduced tension
   e. Water binding affinity is high due to a preponderance of dermatan sulphate in the ECM

2. Zone 2 - The anterior corneal stroma peripheral to the LASIK flap

   a. Tissue tension is markedly reduced
   b. Pif, which is only mildly negative, is driven more negative locally by inflammatory cytokines although likely not to the same level of negativity as in Zone 1 due to the following reasons
      i. Local production, induction and access for inflammatory cytokines in this zone is more limited as the epithelial barrier is generally undisturbed and stromal injury from the mechanism of flap creation and excimer laser photoablation does not occur
      ii. Production and induction of inflammatory cytokines for this zone is predominantly located in the region of the flap sidewall where the epithelial barrier is most disturbed and compromised
      iii. Fluid moves into the zone from
         (a) The limbus and
         (b) Through the flap sidewall during the period of time when tight junctions are disrupted
   c. The slope of the compliance curve is decreased due to reduced tissue tension

d. Collagen interweaving peripherally tends to reduce or inhibit tissue expansion
e. Water binding affinity is high
f. The tissue remains at the same distance from the endothelial pump as existed in the preoperative state

3. Zone 3 – The stromal bed that lies immediately posterior to Zone 1 and 2

   a. Tissue tension is increased, particularly as the tissue absorbs fluid. However, collagen fibrils are inelastic. Most importantly, the local clinical effect of this increase in tension varies between Zone 3a compared to Zone 3b
   b. Pif is driven more negative locally by inflammatory cytokines.
   c. Production and induction of inflammatory cytokines for this zone is predominantly located in the region of the flap sidewall where the epithelial barrier is most disturbed and compromised
   d. Fluid moves into the zone from
      i. The limbus and
      ii. Through the flap sidewall during the period of time when tight junctions are disrupted
   e. The slope of the compliance curve is increased making the tissue less able to imbibe fluid or, if severe, may even cause fluid to be expressed from the ECM as occurs in Zone 3b
   f. Collagen interweaving peripherally tends to reduce or inhibit tissue expansion
   g. Water binding affinity is still higher than in stroma located more posteriorly
   h. The tissue remains at the same distance from the endothelial pump as in the preoperative state. Despite this, with increasing negativity of Pif and fluid influx in the zone, the immediate local effect of the endothelial pump and its ability to keep this zone dehydrated is decreased
   h. The clinical effects of changes in these parameters vary markedly from corneal periphery to center. We have divided effects of this spatial variance into two interrelated sub-compartments – Zone 3a and 3b.

4. Zone 4 - The residual stromal bed posterior to Zone 3

   a. Tissue tension in this zone is likely only nominally changed. If at all, tension may be slightly increased in the more anterior region of Zone 4
   b. Pif is likely not changed significantly over the preoperative state.
   c. The slope of the compliance curve is unchanged
   d. Water binding affinity is lower and the tissue gives up water more readily due to the preponderance of keratan sulphate located in this zone compared to the stroma located more anteriorly
   e. The tissue remains at the same distance from the endothelial pump as in the preoperative state and is most proximate to its immediate effects

Zone 1 - the LASIK Flap Effects of Pif, Reduced Tension and Increased Compliance

In the flap post LASIK, a profound change in tissue performance is created by marked alterations in tissue strain or tension. Corneal tissue tension or load is markedly reduced because virtually all of the collagen fibrils connecting the flap to the residual stroma have been severed by the surgical event. In the absence of tension in the flap, the slope of the compliance curve is decreased causing the incremental change in tissue volume for a specific change in Pif to increase. In addition, the tissue may assume a larger volume, even at a constant Pif. Fortunately, in a normal post-LASIK case these changes are clinically generally transient and insignificant.

> In the LASIK flap, fluid moves into the tissue through breaks in the epithelium and the limbus. As Pif becomes increasingly positive, gross flap edema and thickening occurs creating focal whitening of the cornea, full thickness macrofolds and measurable thickening of the flap.

However, in some cases, inflammatory cytokines produced by epithelial injury or exposure to endotoxins aggressively drives Pif more negative. A highly negative Pif created by such peri-operative surgical trauma causes the corneal flap tissue (Zone 1) to imbibe fluid. Based upon the work of Wiig[166], we know that Pif in the corneal center is markedly more negative than in the periphery. Fluid flowing into Zone 1 from the tear film and limbus will be pulled to the center of the flap by this Pif gradient[167]. However, fluid uptake now follows the new ΔVol / ΔPif relationship, defined by a tissue system functioning under conditions of reduced tension. In this setting, relatively small changes in Pif can contribute to large changes in volume.

Typically, the compliance of the ECM and capacity of the cornea to remove fluid is sufficient to buffer this fluid flux. As a result, these changes do not lead to frank tissue edema that is observable at the slit lamp. A mild increase in the separation of collagen fibrils from an expanding ECM may increase collagen periodicity and account for some of the light dispersing glare, "rainbow glare"[201] or halo effects that patients note in the early days following laser vision correction procedures.

However, in severe cases, as fluid moves into the tissue through breaks in the epithelium and the limbus and Pif becomes increasingly positive, tissue volume centrally may begin to expand at a very rapid rate. The proteoglycan component present in the flap possesses a high degree of affinity for water, further increasing swelling pressures.

If severe, this may cause gross flap edema and thickening. As tissue swelling increases the central tissue where Pif is most negative becomes saturated. This creates focal whitening of the cornea, full thickness macrofolds and measurable thickening of the flap. However, once this central tissue attains saturation, the gradient drawing fluid to the center is substantially reduced. The elimination of the Pif gradient combined with the re-establishment of the barrier function created by tight junctions in the epithelial layer inhibits additional fluid and cell movement towards the corneal center. Over time, with recovery of keratocyte function and normal Pif, the cornea of Zone 1 will regain much more normal fluid dynamics. As a result, tissue clarity and shape will ultimately return to a more normal state. For clarity, it should be noted that in our model there is no significant necrosis of the ECM.

> However, in cases of Stage 4 DLK, CTK, CFN, FNS and CLK it is also critically important to recognize that this loss of tissue transparency, whitening, folding and thickening occurs exclusively in the center of the flap, rather than in a more diffuse manner. This is particularly enigmatic and has served to create much of the confusion in the literature on its etiology

However, in cases of Stage 4 DLK, CTK, CFN, FNS and CLK it is also critically important to recognize that this loss of tissue transparency, whitening, folding and thickening occurs exclusively in the center of the flap, rather than in a more diffuse manner. This is particularly enigmatic and has served to create much of the confusion in the literature on its etiology.

Equally perplexing, this area of severe focal edema is most often broadest at the surface of the cornea just below the epithelium and then progressively narrows as one continues back towards the endothelium creating a "cone-like" area of affected tissue (Figure 11 and Figure 12).

This effect has been described by Hainline et el.[24], as well as being observed in all cases in our series. Hainline et el.[24] attribute this effect the introduction of an inciting substance into the anterior through "microscopic" epithelial defects accompanied by a progressive decline in the affect of a tissue toxin as this unidentified agent advanced from the corneal surface towards the endothelium. They also suggest that a toxin could be introduced into the interface and then become "preferentially attached to the anterior stromal Bowman's area". We believe that these notions to be incorrect as they represent magical thinking and an appeal to a highly implausible hypothesis. Since this "cone" effect only becomes clinically evident several days following an uncomplicated LASIK procedure, the lack of any observable epithelial disruption intraoperatively and postoperatively and delayed timing of its onset makes such an explanation requiring unseen "microscopic epithelial defects" unlikely. Moreover, the notion that a specific toxin can demonstrate a greater affinity for sub-Bowman's stroma compared to other stroma has no scientific basis or precedent.

This "cone shaped" opacification is also particularly challenging if one appeals to the Sonmez and Maloney[22] photoactivation model since, in this case, a photoactivated toxin would likely have its most deleterious affect at the level of the interface and the effect would be expected to decrease as one moved away from the interface in either an anterior or posterior direction. This is not consistent with clinical observations of affected patients. Moreover, to date there is no known toxin, and particularly not Cidex®, that could diffuse through the cornea in such a discrete and localized manner. As such, we posit that these notions represent magical thinking.

> The centrally located cone shaped area of focal edema is caused by properties inherent in the cornea itself. These factors include .the following:
>
> - Anisotropic fluid movement patterns
> - The Pif gradient
> - Fluid trapping compartment syndrome

In our model, the centrally located cone shaped area of focal edema is caused by properties inherent in the cornea itself (Figure 2).

1. **Anisotropic fluid movement patterns:**

Fluid movement from periphery to center in the cornea follows a "fan shaped" pattern. This fan shaped or inverse parabolic fluid trajectory is created by the "onion-like" anisotropic properties of corneal anatomy and structure. Similar to how femtosecond laser pulses follow the path of least resistance, we posit that fluid physically tracks more easily along the lamellar structure of the cornea rather than moving across lamellae. Fluid moves more easily in a direction parallel to layers of collagen mesh than perpendicular to such tightly packed arrays. This pattern of tension modulated fluid flux is well described in the field of hydrogel hydration[202,203]. By analogy, it is much easier to drive a nail "with" the grain of a piece of wood than "across" the grain.

2. **Pif gradient creates a "conveyor belt" effect:**

As fluid moves towards the corneal center, traveling from the periphery to the center, it is progressively pulled towards the posterior of the cornea and is actively pumped out of the center of the cornea due to;

a. The increasing negativity of the Pif gradient in the corneal center and;

b. The local influence of the endothelial pump mechanism as fluid and tissue become progressively more proximal to this pumping process. This serves to overcome fluid layering created by anisotropic characteristics of the tissue

The combination of Pif gradient and anisotropic fluid movement effectively creates a "conveyor belt" effect wherein fluid is moved primarily from the limbus to the corneal center where it is pumped into the anterior chamber. Moreover, this conveyor belt effect causes fluid to follow an inverse parabolic trajectory as it moves from limbus to corneal center. This process also allows for the most efficient nutrient supply to the cornea.

3. Fluid Trapping Compartment Syndrome

A marked mismatch in compliance characteristics occurs between the corneal stromal tissue just posterior to the LASIK flap (Zone 3b) compared to the flap itself (Zone 1). The reasons for this mismatch are discussed in greater detail in a later section of this paper. However, these compliance differences are most intense in the center of the flap. Specifically, compliance in the flap (Zone 1) is increased compared to tissue closer to the endothelial pump in Zone 3b. This compliance mismatch creates an endpoint where fluid is trapped in the center of Zone 1. This fluid trapping or compartment syndrome occurs because the corneal tissue in Zone 3b is under much greater tension than Zone 1 and as a result demonstrates much less fluid compliance or ability to absorb fluid compared to the flap (Zone 1). It also contains proteoglycans that possess significantly less affinity for water than Zone 1. These factors create two physically adjacent compartments that demonstrate markedly different abilities to absorb fluid influx that is moving centrally from the limbus and the area of epithelial disruption near the flap sidewall.

Effectively, it is these differences in the physical properties of Zone 1 and

Zone 3b that cause fluid to be trapped in the center of Zone 1. Excessive tissue compliance mismatch, induced by changes in Pif and tension in Stage 4 DLK, CTK, CFN, FNS and CLK, create an endpoint where fluid cannot easily flow from compartment or Zone 1 (flap) to compartment or Zone 3b (residual stromal tissue just below the flap). In the context of the anisotropic and Pif gradient characteristics of fluid movement in the cornea, this trapping is most intense in the center of the cornea and in the corneal flap thereby creating the cone shaped area of focal edema with its broadest base just below the epithelium and the apex of the cone at the level of the LASIK interface.

As a result of this compliance mismatch, the resulting difference in the fluid clearing performance of Zone 1 and Zone 3b and the trapping of fluid induced by this phenomenon, the central flap turns white with edema and develops macro and microfolds. This makes the flap appear to be "necrotic" while the adjacent stroma appears to be not "necrotic" as was described by Hailine et al.[24] Without question, we agree and our model predicts that the two adjacent tissues appeared to be different in texture. In DLK, CTK, CFN, FNS and CLK these two tissues (flap and adjacent stroma) are, in fact, very physically different and demonstrate markedly different performance characteristics and fluid dynamics. However, these differences are not caused by a necrotic process that is somehow restrained or contained exclusively in the flap while adjacent stroma is unaffected, since no known toxic or necrotic process exists which is able to create such a biological event.

Our model also posits that Interface Fluid Syndrome (IFS)[204-208] associated with DLK treatment and elevated intraocular pressure (IOP) occurs because of an exacerbation of this compartment syndrome that is caused by IOP elevation from topical steroid use. The increased IOP worsens the compliance mismatch and fluid trapping problem by increasing corneal tension in Zone 3b through tissue compression. It also reduces reserve capacity for the endothelial pump by driving fluid from the aqueous

Based on our model of corneal fluid dynamics, Pif gradients and the interaction between tension and tissue compliance, severe local and focal flap edema, cone shaped cornea flap whitening, stromal macrofolds and tissue opacification will consistently and exclusively occur in the central and paracentral cornea.

compartment into the stroma[208]. These factors make it increasingly difficult for fluid collecting in the flap to move along normal anisotropic fan shaped trajectories into the residual stromal bed. Excess fluid thereby accumulates based on 'regions' of least resistance first in the central epithelium in the form of epithelial bullae followed by collection in the interface. Once the IOP is reduced the tissue compression and tension in Zone 3b is also reduced and fluid may begin to move more readily between these adjacent compartments and the IFS will resolve.

As a consequence, based on our model of corneal fluid dynamics, Pif gradients and the interaction between tension and tissue compliance, severe local and focal flap edema, cone shaped cornea flap whitening, stromal macrofolds and tissue opacification will consistently and exclusively occur in the central and paracentral cornea.

Irrespective of the complexity of the physical factors described in our edema model (we apologize that the cornea is not of a more simple design), we vigorously assert that tissue necrosis in general and flap necrosis in particular are simply non-viable explanations for a consistently reversible event in the central cornea

Unfortunately, prior to the introduction of our model, these features have been commonly considered to occur as a consequence of tissue necrosis. Based on the absence of scientific evidence in favor of tissue necrosis combined with a compelling argument against such an event, we posit that the necrosis paradigm is highly erroneous. On the contrary, focal corneal edema is a reversible process that, with appropriate management, can recover over time without scarring or visual compromise. We suggest that this physiologic process is better aligned with the clinical picture than a reliance on tissue necrosis to explain these events. Irrespective of the complexity of the physical factors described in our edema model (we apologize that the cornea is not of a more simple design), we vigorously assert that tissue necrosis in general and flap necrosis in particular are simply non-viable explanations for a consistently reversible event in the central cornea.

Zone 2
Effects of Pif, Tension and Compliance

In the peripheral anterior zone of the cornea (Zone 2) following excimer laser vision correction, Pif becomes more negative, but not as negative as found in Zone 1. Similar to Zone 1, tension on collagen fibrils is significantly decreased. This loss of tension caused by severing of collagen fibrils creates a more compliant system that can hold more fluid than in its preoperative condition.

> In Zone 2, tissue swelling creates a marked transient peripheral steepening. This steepening contributes to a reversible hyperopic shift and the induction of positive spherical aberration

The increased compliance of the peripheral anterior stromal compartment (Zone 2) allows this tissue to swell and creates a thickening of the local corneal tissue. In contrast to Zone 3, this tissue swelling has no significant impact on tissue strain. However, from a refractive perspective, this tissue swelling creates a marked transient peripheral steepening. This steepening contributes to a hyperopic shift and the induction of positive spherical aberration. As Pif becomes more normal, this peripheral swelling and steepening demonstrates reversibility.

Zone 3 - Residual Bed
Effects of Pif, Tension and Compliance

Zone 3 involves tissue in the middle region of the cornea effectively "sandwiched" between Zone 1 and 2 above and Zone 4 below. In addition, based on tissue response characteristics, two additional distinct sub-zones (Zone 3a and 3b) are created in Zone 3. Zone 3a is located in the peripheral cornea and is defined by a circular area that begins adjacent to the vertical sidewall of the LASIK flap and extends to the limbus. It is located just below Zone 2. Zone 3b defines the residual stromal bed just below the LASIK flap (Zone 1).

> In Zone 3, alterations in Pif, fluid influx, tension and tissue compliance cause two related clinical effects;
>
> 1. The reversible hyperopic shift. The contribution of Zone 3 is significantly greater than that of Zone 2.
> 2. A compartment syndrome that traps fluid in the central flap (Zone 1)

Clinical Effects

Similar to Zone 1 and 2, the pathologic processes in Zone 3 are all fundamentally driven by changes in Pif. However, in Zone 3, the dynamic relationship between mechanical tension and tissue swelling due to these changes in Pif is far more complex. Alterations in Pif, fluid influx, tension and tissue compliance in Zone 3 cause two related clinical effects;

1. The reversible hyperopic shift. The contribution of Zone 3 is significantly greater than that of Zone 2.
2. A compartment syndrome that traps fluid in the central flap (Zone 1) previously described in this paper

Cause of Clinical Effects

Similar to Zone 1 and 2, inflammatory cytokines and other factors drive Pif negative in Zone 3. However, tissue tension dynamics are markedly different in Zone 3 compared to Zone 1 and 2 and the effect of tension on tissue compliance is of major clinical significance. Firstly, injured epithelium surrounding the flap sidewall typically represents the largest source of inflammatory cytokines in the cornea and the local disruption in the normal barrier function becomes the primary location for diffusion of these inflammatory cytokines into the local stroma. Secondly, this area of the flap sidewall is also the path of least resistance for fluid and inflammatory cells from the tear film. Thirdly, fluid in the tear film that is drawn into the interface is rapidly absorbed by the corneal stroma near the flap sidewall due to its highly negative Pif caused by these local cytokines. In addition, the increase in negativity in Pif in the flap sidewall region begins to draw significant amounts of fluid from the limbus. As fluid is selectively absorbed by stroma near this circular entry portal, the ECM begins to expand and the tissue swells and thickens. Fluid flowing into the cornea from the limbus and the entrance portal created by the flap

sidewall in response to the local decrease in Pif in the area combine to create a preponderance of swelling and thickening in Zone 3a compared to Zone 3b.

> As the tissue in Zone 3a swells it places progressively more tension on the collagen fibrils in Zone 3b. Without an ability to increase their cord length, this increased tension causes marked compression and a severe flattening effect in Zone 3b. At the same time, this swelling creates a localized steepening in Zone 3a.

As the stroma in Zone 3a becomes increasingly edematous the cornea in that region begins to expand and thicken. Corneal edema in Zone 3a causes the collagen fibrils present there to move further apart. However, the eye is a closed hydraulic system and it is effectively incompressible. As a consequence, as the collagen fibrils move further apart, they expand proceeding outwards away from the center of the eye in a centrifugal direction. During this tissue expansion process, the origins of these collagen fibrils remain fixed in position and tethered at the limbus and cannot move. Since collagen is not elastic, this peripheral outward tissue expansion with fixed anchor points at the limbus begins to place increasing tension on those collagen fibrils that remain intact in the central cornea. In order for tissue expansion to occur in a uniform symmetrical manner in Zone 3 the cord length of the collagen fibrils would need to increase, particularly in the fibrils in the outer aspect of Zone 3. An increase in cord length is not possible. As a consequence, the curve created by these collagen fibrils becomes distorted such that they are steepened and flattened in a localized fashion. As the tissue in Zone 3a swells it places progressively more tension on the collagen fibrils in Zone 3b. Without an ability to increase their cord length, this increased tension causes marked compression and a severe flattening effect in Zone 3b. At the same time, this swelling creates a localized steepening in Zone 3a. This process of central compression and flattening is likely enhanced by the local capacity of the endothelium to dehydrate Zone 3b more easily than Zone 3a. The resulting imbalance occurs rapidly, much like would occur if unequal weight is applied to a balance beam scale. As weight is added to one side and simultaneously taken away from the opposing end, the balance beam tips rapidly. The same imbalance in ECM tissue volume in the cornea results in rapid expansion of the mid-peripheral cornea and compensatory compression of the central residual stromal bed.

> A simple analogy is the distortion of a dome shaped camping tent if outward pressure is applied. As the mid peripheral area of the dome tent moves outwards, the center and apex of the dome tent must flatten in a compensatory fashion. This effect results in local peripheral steepening of the dome tent walls combined with central flattening of the dome of the tent

A simple analogy to assist in understanding such a phenomenon might be to envision the distortion of a dome shaped camping tent if outward pressure is applied to the mid periphery. Similar to the cornea, adding to the cord length of the nylon fabric is not possible and the base of the tent is tethered to the ground and cannot move. Therefore, as the mid peripheral area of the dome tent moves outwards, the center and apex of the dome tent must flatten in a compensatory fashion. This effect results in local peripheral steepening of the dome tent walls combined with central flattening of the dome of the tent. In the cornea, these same forces also result in mid-peripheral steepening and central flattening. For reference, this is highly similar to the phenomenon described in the Roberts biomechanical model. However, in our model, this effect is created by biological events and is largely reversible when those biological changes begin to trend back towards normal.

This peripheral steepening and central flattening may also be illustrated in the following manner. Imagine a string fixed at each end with a push pin. Such a string extending between these two push pins can be positioned to form a semi-circle with a constant radius. We can then divide the length of the curved string into 4 sections by placing a label at each end of the string (points A and E) and at points that will divide the overall length of the string into three equal segments (B,C,D). If we then take two pencils and place one pencil at point B and the other at point D and attempt to move the string away from the centroid of its radius, the semi-circle will become distorted and will no longer remain spherical in its contour. Outward movement of points B and D will result in the section of string extending between points B and D becoming straightened (or flattened). Simultaneously, the local radius of curvature of the short length of string located around points B and D will be increased (or locally steepened). Again, this is the identical type of reversible deformation that occurs in the cornea. For reference, points B and D correspond to the edge of the LASIK flap and the section of the string between points B and D corresponds to the collagen fibrils found in Zone 3b. The section of string between points A and B and between D and E similarly correspond to the collagen fibrils found peripherally in Zone 3a and demonstrate localized flattening.

> The hyperopic shift is caused by biologically mediated central corneal flattening, mid-peripheral steepening and peripheral flattening. Ultimately, the swelling of Zone 3a is responsible for inducing local peripheral steepening. Conversely, localized thickening of Zone 3a also causes the flattening and thinning of Zone 3b

Hyperopic Shift

The hyperopic shift in cases of Stage 4 DLK, CTK, CFN, FNS and CLK are caused by this biologically mediated central corneal flattening, mid-peripheral steepening and peripheral flattening. Ultimately, the swelling of Zone 3a is responsible for inducing local peripheral steepening. Conversely, localized

thickening of Zone 3a also causes the flattening and thinning of Zone 3b. The reason that the "knee" or "fulcrum" of this swelling in Zone 3a occurs in the mid-periphery is due to the collagen interweaving that occurs in the peripheral corona of the cornea and the rigid tethering of collagen fibrils to the limbus. These factors restrict expansion of the most peripheral aspect of Zone 3a and creates disproportionate swelling centrally compared to peripherally. An analogy would be to conceive of Zone 3a as a balloon in which one ½ of the balloon has been covered with an adhesive tape. If such a balloon is inflated (becomes edematous) the expansion will be asymmetrical with the majority of the expansion occurring in the ½ of the balloon that is not restricted by the tape.

This asymmetric edema effect occurs to some degree in most cases of myopic LASIK. However, in cases of DLK, CTK, CFN, FNS and CLK the effect can be extremely large, as seen in the OCT and Pentacam™ images (Figures 9 through 11). In virtually all cases, the central flattening, mid-peripheral steepening and peripheral flattening effect is reversed when Pif and fluid dynamics become normalized. It is for this reason that the hyperopic shift in these disorders is reversible.

Compartment Syndrome

Increasing tissue tension has the effect of increasing the slope of the compliance curve and decreasing tissue compliance. As a result of compression and flattening of the central cornea (Zone 3b) the compliance of the Zone 3b compartment is decreased. With increasing collagen tension and decreasing tissue compliance, fluid is actually excluded from the tissue following a new ΔVol / ΔPif compliance curve that is shifted to the left of normal. Due to this change in tissue compliance, the residual corneal bed in the center of Zone 3b becomes transiently thinner. This thinning is visible at the slit-lamp. Moreover, this thinning of the tissue due to tissue compression further contributes to the marked central flattening and exacerbates the hyperopic shift (Figure 11).

As tissue compliance decreases in Zone 3b, the ability for fluid to move from the flap (Zone 1) into Zone 3b decreases. If severe, the change in compliance creates a compartment block or syndrome wherein fluid moving into Zone 1 becomes trapped and is unable to move into Zone 3b. This effect is further compounded by the high affinity for water by the GAG's in Zone 1 combined with the increased tissue compliance in Zone 1. It is this process that causes the central flap to become extremely edematous causing focal central flap whitening, a "cone shaped" white area as described previously, flap macrofolds and tissue opacification (Figure 12).

> Hyperopic shift is caused by transient central corneal flattening, mid-peripheral corneal steepening and peripheral flattening. Compliance mismatch between corneal components contribute significantly to a compartment block that traps fluid in the center of the flap.

The fan shaped inverse parabolic fluid movement pattern created by the anisotropic structure of the cornea determines that this whitening and flap folding occur directly in the corneal center (Figure 2). However, as inflammation and other factors subside in the postoperative period, this compartment block gradually resolves and the central flap whitening and macrofolds slowly fade. There is no clinically relevant necrosis or stromal remodeling. If the patient exhibits loss of BSCVA it is a result of unresolved macrofolds in the flap. Residual microfolds that adversely affect BSCVA must be corrected surgically or visual recovery will be limited or markedly delayed.

Our model predicts that these biomechanical effects will be worse in high myopic cases affected by Stage 4 DLK, CTK, CFN, FNS and CLK than in low myopia, although other factors that influence tissue tension such as corneal thickness can occasionally appear to confound this tendency. This is corroborated by the observation that the two most seriously affected patients in our series (Case #1 and #3) were the most myopic patients. Hainline et al.[24] also noted that myopic patients were more affected in their case series as was also true in the series published by Lyle and Jin[26].

> The mechanical processes in Zone 3 are similar to the Roberts model.. However, in our model changes in mechanical forces substantively result from biological changes determined by control of loss of control of local Pif rather than loss of collagen fibrils.

Summary

In summary, transient central corneal flattening, mid-peripheral corneal steepening and peripheral flattening cause the hyperopic shift observed in cases of Stage 4 DLK, CTK, CFN, FNS and CLK. They also contribute significantly to a compartment block that traps fluid in the center of the flap. It should be noted that the mechanical processes described in Zone 3 of our model are similar to those effects described by Dr. Roberts[184-186].

However, rather than occurring almost entirely on the basis of mechanical change in collagen tension due to reduction in numbers of intact collagen fibrils, our model suggests that these changes in mechanical forces substantively result from biological changes determined by control of loss of control of local Pif. Since changes induced in control of Pif from refractive surgery can be reversed with tissue healing, the hyperopic shift and central corneal edema so induced also exhibit features of reversibility. Moreover, the reversibility of the effect is clearly documented in the Pentacam™ tomography and OCT studies included in this paper (Figure 9 through 11).

Zone 4
Effects of Pif, Tension and Compliance

In Zone 4, which is located just posterior to Zone 3, the cornea is not significantly impacted by events of the surgery. Overall tissue tension in Zone 4 is not much different than in the pre-surgical state, or at worst, is slightly increased in its more anterior region. In addition, the effect of inflammatory cytokines is less than in more anterior areas of the cornea. As a result of nominal changes in tissue tension and Pif, the slope of the compliance curve remains relatively unchanged. The proximity of the tissue to the endothelial pump is also unchanged from the pre-operative state. This latter observation, combined with the low water binding characteristic of keratan sulphate located most posteriorly in the cornea, allows this area of the stroma to give up its water more readily than in Zone 1, 2 or 3.

> In Zone 4, the cornea is not significantly impacted by events of the surgery. A s a result, Zone 4 does not contribute significantly to the clinical manifestation of DLK, Stage 4 DLK, CTK, CFN, FNS or CLK

As a result, Zone 4 does not contribute significantly to the clinical manifestation of DLK, Stage 4 DLK, CTK, CFN, FNS or CLK.

Reversibility of the Hyperopic Shift

As previously noted, the topographic and refractive effects described in our disease model for DLK, Stage 4 DLK, CTK, CFN, FNS and CLK are similar to the biomechanical model proposed by Roberts[184-186]. However, the Roberts model does not provide any significant mechanism for reversibility because it generally discounts biological events and overemphasizes the influence of mechanical forces created by loss of collagen fibrils. As a result, the Roberts

> The Roberts biomechanical model does not provide any mechanism for reversibility because it discounts biological events and overemphasizes the influence of mechanical forces created by loss of collagen fibrils. As a result, the Roberts model is too simplistic and does not accurately represent the physiologic process actually occurring in the cornea.

model is too simplistic to be a value in describing the actual interplay between biological events and mechanical forces in these disorders and does not accurately represent the physiologic process actually occurring in the cornea. Our model provides for an explanation of the biomechanical events in a setting where reversibility of effect is not only possible, it is fully anticipated and predicted.

In our model, as inflammatory cytokines wane and the keratocyte population physiology recovers over time, control over Pif is re-established and the tendency of the tissue to imbibe fluid is markedly diminished. As Pif dynamics become increasingly more normal, the compliance of the three zones in the cornea begin to drift towards a more homeostatic state. The result of this normalization in control of Pif is that the hyperopic shift gradually resolves over time. Swelling pressures that create collagen tension centrally and peripherally become more normal. Depending upon the physiologic response of the specific cornea, the hyperopic shift may resolve completely or may leave the patient with a residual hyperopic result. If an induced astigmatism was observed due to asymmetry in these forces, that astigmatism will generally resolve as well.

> Our model provides for an explanation of the biomechanical events in a setting where reversibility of effect is not only possible, it is fully anticipated and predicted

Although this paper does not provide the opportunity to discuss this in detail, it should be becoming increasingly obvious that these same biomechanical events occur to some degree in most cases of laser refractive surgery. They create transient refractive overshoots and undershoots that heretofore have had no physiologic basis or explanation. The model also assists in explaining such phenomenon as our ability to modulate the refractive endpoint of PRK surgery by titrating the dosing of topical steroids, why refractive endpoints with LASIK or PRK are more accurate in newer scanning laser systems that reduce thermal and acoustic load to the cornea and why a significant reduction in pulse energy in femtosecond lasers produce much less transient hyperopic overshoot and DLK in Intralase™ cases (Will BR. Strategies for Optimizing Flap Excision and Bubble Formation. IntraLase Users Meeting, Dana Point, California, June, 2002).

Interface Fluid Syndrome and the Effects of Elevated IOP

Interface fluid syndrome is currently considered to be a distinctly different entity from DLK or CTK[208]. We suggest that this belief is in error. Rather, IFS is caused by the same mechanisms that cause compartment block in Stage 4 DLK, CTK, CFN, FNS and CLK. Because the elevation of IOP is most often related to steroid dosing for the treatment of DLK (or associated entities), IFS generally represents a progression or end stage of the same biologically mediated mechanical events that occur in DLK, Stage 4 DLK, CTK, CFN, FNS and CLK.

Our model posits that IFS occurs because of an exacerbation of the compartment syndrome observed in Stage 4 DLK, CTK, CFN, FNS and CLK. The exacerbation is caused by the elevation in IOP induced by topical steroid use. The increased IOP worsens the compliance mismatch and fluid trapping problem because it increases corneal tension in Zone 3b. It also reduces reserve capacity for the endothelial pump by driving fluid from the aqueous compartment into the stroma. These factors make it increasingly difficult for fluid that is

collecting in the flap (Zone 1) to move along normal anisotropic fan shaped trajectories into the residual stromal bed (Zone 3b). Excess fluid thereby accumulates first in the central epithelium in the form of epithelial bullae and then begins to collects in the interface. Because the increased tension in Zone 3b and compartment block in IFS is primarily caused by the elevated IOP, IFS can occur in the absence of interface cells and inflammatory cytokines and in the absence of central focal edema of the flap. However, aggressive dosing with high dose topical or oral steroids is typically associated with treatment of DLK in the post LASIK cornea. As a result, the processes are commonly linked.

IFS is caused by a compliance mismatch that causes fluid trapping in the corneal flap and LASIK interface induced by an elevated IOP. IFS and DLK are commonly associated because elevation of IOP in the post LASIK eye is almost exclusively due to aggressive dosing of topical or oral steroids. However, our model also predicts that IFS will occur in a post LASIK eye that exhibits elevated IOP from any cause and may occur independently from DLK

As a consequence, we believe that IFS is a "compartment" syndrome and represents an pressure induced exacerbation of the difference in tissue compliance between Zone 1 and Zone 3b. The disease mechanism is essentially identical to the cause of flap edema in Stage 4 DLK as previously described in this paper. As such, one could reasonably consider IFS as "Stage 5 DLK", although we vigorously reject the Linebarger et al.[4] staging paradigm. It may also occur independently from DLK if IOP becomes elevated due to other mechanisms. Based on this mechanism, IFS may also predictably occur in the absence of DLK, particularly if the eye is being treated with topical steroid medications.

The cornea can generally quickly recover from IFS. Once steroids are discontinued and the IOP becomes normal, the tension in Zone 3b is also reduced. With reduction in tension the compliance mismatch and fluid trapping is improved and fluid may begin to move more readily between these adjacent compartments. Fluid in the interface and flap can then move into the residual stromal bed (Zone 3b) and be pumped into the anterior chamber by the endothelial pump mechanism.

In summary, a compliance mismatch that causes fluid trapping in the corneal flap and LASIK interface induced by an elevated IOP is the primary mechanism causing IFS. The mechanisms causing IFS are essentially identical to those that cause Stage 4 DLK, CTK, CFN, FNS and CLK. These disorders are commonly associated because elevation of IOP in the post LASIK eye is almost exclusively due to aggressive dosing of topical or oral steroids for treatment of inflammation. However, our model also predicts that IFS will occur in a post LASIK eye that exhibits elevated IOP from any cause. As a consequence, it may occur independently from DLK.

General Issues

Accuracy of the Model

Model Predictions

An attribute of a good clinical model is its ability to make accurate predictions regarding responses of the cornea to a

> Our model provides both accurate predictions and significant insight into the behavior of the cornea when subjected to a number of surgical and non-surgical processes

variety of clinical events. We believe that, in addition to explaining the pathophysiology of DLK, Stage 4 DLK, CTK, CFN, FNS and CLK, this model provides both accurate predictions and significant insight into the behavior of the cornea when subjected to a number of surgical and non-surgical processes.

Myopic Shift with Corneal Dehydration

Our model predicts that excessive corneal hydration following LASIK leads to a reversible hyperopic shift caused by central corneal flattening, mid-peripheral corneal steepening and peripheral flattening. This is the pathophysiologic cascade seen in DLK, Stage 4 DLK, CTK, CFN, FNS and CLK. However, conversely, our model also predicts that

excessive corneal dehydration following LASIK will create a reversible myopic shift characterized by central corneal steepening, mid-peripheral flattening and peripheral steepening.

Fam et al.[209] describe an ethnic Chinese adventurer who experienced a visually significant myopic shift in both eyes following 2 weeks at the geographic North Pole where he experienced 0% humidity, -50°C temperatures and 55 km/h winds. One week after return from this trip his right eye exhibited -2.50 – 0.50 X 180 and his left eye -1.50 – 0.75 X 10 with no loss of BCVA. Three months after return from the expedition, the

> Our model predicts that excessive corneal dehydration following LASIK will create a reversible myopic shift characterized by central corneal steepening, mid-peripheral flattening and peripheral steepening. This phenomenon has been observed and documented in post LASIK patients subjected to severely dry environmental conditions. Our model provides a physiologic basis for the reversibility of this myopic shift

patient's refraction spontaneously improved to pre-expedition levels of

Plano -0.50 X 170 in the right and Plano in the left eye.

Serial Orbscan® II measurements demonstrated central corneal steepening coupled with peripheral flattening that developed during the North Pole trek as the cause for the myopic shift. As the myopic shift resolved the central cornea became flatter and the peripheral cornea became progressively steeper. No change was observed in the posterior corneal curvature.

Consistent with the predictions of our model, the authors correctly theorized that with corneal desiccation, the central and peripheral areas of the cornea behave in an opposing manner. Specifically, as the corneal periphery becomes thicker the corneal center becomes flatter and as the center becomes steeper the periphery becomes flatter. They postulated that as the cornea desiccated during the trek that the peripheral thickening and central flattening of the cornea created by the initial LASIK procedure was temporarily reversed.

White and Mader[210] as well as Boes et al.[211] have reported myopic shifts in refraction in LASIK patients subjected to high altitudes and extremely dry environments. In such cases the authors speculated that such shifts were caused by hypoxia induced endothelial dysfunction, a mechanical effect of severing anterior stromal fibers in an eye exposed to low atmospheric pressure or a non-uniform increase in corneal thickness of unknown etiology. Although corneal hypoxia could alter corneal cell behavior, we suggest that these myopic shifts are more likely related to corneal desiccation accompanied by reversible central corneal steepening and peripheral flattening.

Our model predicts a myopic shift associated with corneal dehydration and provides a physiologic basis for the reversibility of this phenomenon. Our model also predicts that the physiologic basis for this myopic shift during corneal desiccation is essentially identical to that observed in corneas recovering from Stage 4 DLK, CTK, CFN, FNS and CLK. It is a biologically mediated reversible biomechanical event governed by changes in Pif.

Anterior corneal thickening and posterior thinning in response to inflammation

Our model predicts that the introduction of inflammatory cytokines such as

> Our model predicts that inflammatory cytokines will cause thickening of the anterior stroma and thinning of the middle and posterior stroma, when the tissue response is observed in the central aspect of the cornea. These predictions are consistent with experimental observations of rabbit corneas which demonstrate a highly localized preferential swelling and thickening of the anterior cornea combined with marked thinning of the posterior cornea on exposure to epithelial injury

prostaglandins and interleukins into the ocular environment will cause thickening of the anterior stroma and thinning of the middle and posterior stroma, when the tissue response is observed in the central aspect of the cornea.

These mechanical processes predicted in our model are consistent with that observed by Ruberti et al.[163] and Karon and Klyce[164] wherein they observed corneal swelling in the presence of epithelial trauma in rabbit eyes to occur in highly localized manner creating a preferential swelling and thickening of the anterior cornea combined with marked thinning of the posterior cornea. The authors attributed this change in tissue thickness and fluid flux to increased oncotic forces created by release of osmotically active macromolecules from apoptotic keratocytes into the stroma.

> Our model predicts that the observed effect of anterior corneal swelling and posterior cornea thinning occurs as a consequence of the effects of inflammatory cytokines on Pif and corneal biomechanics rather than the osmotic effects of keratocyte apoptosis as has been previously suggested as a possible mechanism

Although we agree with the author's observations, we assert that the mechanism suggested by the authors to explain this phenomenon is in error. We believe, and our model predicts, that the observed effect occurs as a consequence of the effects of inflammatory cytokines on Pif and corneal biomechanics as we have described in our model for DLK. Essentially, inflammatory cytokines diffusing into the corneal stroma

following epithelial removal initiate a sequence of events that cause increased tension in Zone 3 of the normal cornea. This results in changes in tissue compliance that result in a process of anterior tissue swelling combined with posterior tissue thinning similar in principle to that observed in the cornea in Stage 4 DLK.

In addition, we believe the speculation by Ruberti et al.[163] and Karon and Klyce[164] that this effect results from oncotic effect of cell apoptosis is contradicted by generally understood mechanisms of cell apoptosis and the specific physiologic events that occur during apoptotic cascades. Contrary to the suggestion of the authors that large osmotically active molecules are acutely dispersed into the ECM by apoptotic keratocytes, in typical apoptotic cascades cytosolic macromolecules are preferentially sequestered until phagocytosis occurs rather than being released in the adjacent tissue to create an osmotic effect[212]. In order for these author's assertion to be true, apoptosis of keratocytes would need to occur in a highly unique fashion compared to other cells in the body. We posit that this latter notion is unlikely to be scientifically defensible.

In addition, although not specifically tested by the authors, our model would also predict that anterior swelling and posterior thinning of the cornea exposed to inflammatory cytokines is a reversible phenomenon. Simple epithelial injury or scratch in the human cornea virtually never results in permanent change in cornea shape. This observation appears to support the notion of reversibility. We posit that it is reasonable to suspect that the process of anterior swelling combined with posterior thinning of the cornea induced by inflammatory cytokines is a rapidly reversible biologically mediated biomechanical event rather than a process caused by osmotic forces.

Model Predicts that Keratocyte Dropout will demonstrate Adverse Effects on Flap Adhesion

Poor flap adhesion following LASIK appears to increase the risk for flap slippage and epithelial ingrowth. Although not typically vision threatening, such complications are nevertheless vexing to both the surgeon and the patient.

> Pif is the single most important force holding the LASIK flap in position immediately following a LASIK procedure. Our model suggests that keratocyte function is of critical important in the control and maintenance of corneal Pif

We believe that Pif is the single most important force holding the LASIK flap in position in the hours and days immediately following a LASIK procedure. Our model suggests that keratocyte function is of critical important in the control and maintenance of corneal Pif. As a result, we predict that a significant decrease in the keratocyte population density in the LASIK flap or corneal stroma in the retroablation area will reduce the ability of the cornea to hold the flap in position following a

LASIK retreatment procedure due to a decrease in the cornea's ability to maintain a highly negative Pif when exposed to the stress of excimer laser surgery.

> The significant decrease in the keratocyte population density in the LASIK flap and stromal retroablation area reduces the ability of the cornea to hold the flap in position following a LASIK retreatment procedure and contributes to the increased risk for epithelial ingrowth in those cases

It is general knowledge amongst experienced LASIK surgeons that the incidence of epithelial ingrowth increases significantly in LASIK retreatments performed years after the initial LASIK procedure. This increased complication rate causes many refractive surgeons to choose PRK as a preferred method of enhancement for LASIK cases more than a few years old. Epithelial ingrowth also occurs far more frequently in cases that were initially completed using a microkeratome versus that observed following Femtosecond laser, irrespective of number of years between the primary or subsequent retreatment procedures. To date, to our knowledge no mechanism has been suggested to explain this phenomenon.

Erie et al.[213,214] measured keratocyte density in the LASIK flap and retroablation stroma using confocal microscopy. They observed that flap keratocyte density was reduced by 22% by 6 months and 37% at 5 years. In addition, stroma keratocyte density in the retroablation zone was reduced by 18% by 1 year and was markedly reduced to 43% at 5 years.

> The marked increase in incidence of epithelial ingrowth observed years after an initial LASIK procedure is directly related to the loss of keratocytes and a concomitant decrease in the cornea's ability to maintain a highly negative Pif when stressed surgically. With a reduced ability to provide a highly negative Pif, flap adhesion is decreased allowing for increased flap edge movement during blinking, migration of epithelial cells at the flap margin and the induction of epithelial ingrowth

Our model predicts that:

1. Epithelial ingrowth rates will measurably increase with increasing age of the flap due to keratocyte dropout and the deleterious effect of the latter on the cornea's ability to maintain a negative Pif and hold the flap securely onto the residual stromal bed following retreatment surgery

2. Microkeratome flaps will be more prone to this effect due to the extremely thin residual stroma located at the edge of the flap compared to a planar Femtosecond flap created using a perpendicular flap edge

3. Use of a combination of hyperosmotic agents such as topical glycerin and the dehydrating effect of a bandage contact lenses will serve to reduce the risk of epithelial ingrowth by

enhancing flap adhesion in the immediate postoperative period

We assert that the marked increase in incidence of epithelial ingrowth observed years after an initial microkeratome LASIK procedure is directly related to the loss of keratocytes in the retroablation zone and the LASIK flap and a concomitant decrease in the cornea's ability to maintain a highly negative Pif when stressed surgically. With reduced ability to provide a highly negative Pif, flap adhesion is decreased allowing for increased flap edge movement during blinking. Decreased flap adhesion provides a track for migration of epithelial cells at the flap margin and the induction of epithelial ingrowth.

> Maintaining a negative Pif through the use of a potent topical hyperosmotic agent (Glycerin) and the dehydrating effect of a soft bandage contact lens markedly reduces the risk and incidence of epithelial ingrowth. Sustaining a highly negative Pif can largely reverse the deleterious effects of keratocyte dropout on the ability of the cornea to hold the flap securely in position during healing

This effect is of particular significance in microkeratome cases since the flap edge architecture is markedly different than that found with an Intralase™ femtosecond device. Keratocyte dropout creates a significantly greater effect in the highly tapered edge of a meniscus flap where keratocyte dropout and scarring is proportionally more critical than in the thick edge provided by the perpendicular edge of a planar femtosecond flap.

In addition, in clinical practice we find that the combined use of a potent topical hyperosmotic agent (Glycerin Preservative free, Leiter's Pharmacy, San Jose, CA, USA) and the dehydrating effect of a soft bandage contact lens markedly reduce the risk and incidence of epithelial ingrowth, particularly when such procedures are performed years following the initial LASIK procedure. In our experience, enhancing the ability of the cornea to maintain a negative Pif and improving flap adhesion acutely following flap lift for LASIK retreatment is a valuable method directed at reducing the risk of epithelial ingrowth in such cases. We believe that supplementing and amplifying the cornea's capacity to sustain a negative Pif through the use of such exogenous dehydrating forces can largely reverse the deleterious effects of keratocyte dropout on the ability of the cornea to maintain a highly negative Pif during surgical stress.

We suggest that clinical observations of increased rates of epithelial ingrowth associated with declining population density of keratocytes in the flap and retroablation area are consistent with predictions derived from our model for DLK, Stage 4 DLK, CTK, CFN, FNS and CLK. We posit that these clinical results increase confidence in the accuracy of our model.

Pharmacologic cytokine analogs will adversely affect Pif, flap adhesion and refractive endpoints

Our model predicts that pharmacologic agents that adversely affect Pif will have unfavorable effects on flap adhesion and the predictability of refractive endpoints. Drugs that increase the production of inflammatory cytokines or alter the cell cytoskeleton by affecting intracellular cAMP levels mediated by G-protein effects will change Pif.

> Our model predicts that pharmacologic agents that adversely affect Pif, such as phenylephrine and Alphagan®, will have unfavorable effects on flap adhesion and the predictability of refractive endpoints. These drugs have their deleterious effects on Pif by causing an increased production of inflammatory cytokines or by altering the cell cytoskeleton by affecting intracellular cAMP levels mediated by G-protein effects

Topical phenylephrine has been linked to flap slippage and poor flap adhesion[43]. We posit that phenylephrine and other potent alpha-agonists cause these complications by interfering with the control of Pif in the cornea following LASIK surgery.

The drug phenylephrine has demonstrated the potential to adversely affect surgical outcomes in ocular surgery and is well known to cause corneal edema[215-219]. These adverse outcomes likely occur as a result of either direct or indirect action of this drug on the control of the fluid dynamics in the cornea. We believe that phenylephrine initiates a cascade that causes corneal edema by binding to α-receptors located on the conjunctival and corneal epithelium, stromal keratocytes and endothelial cells.

Although the clinical effect depends somewhat on the specific cell type affected, we suggest that the primary effect of this binding is that it causes epithelial cells to activate phospholipase A_2, which converts membrane phospholipids to arachidonic acid. Activity by the cyclooxygenase pathway generates prostaglandins that exert their effects locally. Prostaglandins produced in this manner may bind to G-protein coupled receptors of keratocytes. This can in turn activate adenylate cyclase and catalyze the production of the second messenger cAMP. Elevated levels of cAMP intracellularly may alter the cellular cytoskeleton of the keratocyte by interfering with the polymerization of F-actin. Interfering with F-actin may result in reduced cytoskeleton initiated tension on beta-integrins and reduced tension on the ECM[42,220-230]. This could drive Pif negative and cause local tissue edema.

Phenylephrine shares a common attribute with Brimonidine tartrate 0.2% (Alphagan®; Allergan, Inc., Irvine, CA) another potent α-agonist in that both drugs contribute to flap dislocation following LASIK. Walter and Gilbert[231] reported a significant increase in both flap dislocation and enhancement rate after LASIK associated with the use of one drop of Brimonidine instilled into the eye minutes before surgery. In their study, the overall incidence of flap folds or dislocations was 0.0% in the non-Brimonidine groups versus 15.4% in the Brimonidine group (P=0.00001). In addition, the enhancement rate was significantly higher (P=0.0004) in the Brimonidine group at 36% compared to the non-Brimonidine groups measured at 9% and 14% respectively.

Our model predicts that drugs such as phenylephrine and Alphagan® will affect flap adhesion and refractive predictability and that this effect will occur as a result of their interference with the control of Pif. It should also be noted that, in the Walter and Gilbert study, patients received only one drop of Brimonidine tartrate 0.2% preoperatively, thereby demonstrating the exquisite sensitivity of the fluid dynamics in the human cornea to α-agonists. Although the authors suggested other possible mechanisms for their observations including ischemia of the anterior segment or an adverse effect of the drug vehicle on flap adhesion, none of these speculations are based upon any known clinical or basic science data.

We propose that well known effects of α-agonists cause flap slippage and refractive inaccuracies. These chemokines increase local production of inflammatory cytokines that bind to G-protein coupled receptors on keratocytes, alter tension on the cytoskeleton and integrins by affecting cAMP levels and actin polymerization and change Pif. Refractive inaccuracies occur as a result of biomechanical influences of

Pif alterations on corneal shape and power as described in our model.

Central islands, Erratic Outcomes and the LADARVision® Device

Second only to the scientific chaos surrounding the DLK, Stage 4 DLK, CTK, CFN, FNS and CLK debacle, perhaps no other complication of refractive surgery is as poorly understood as the problem of central islands. Despite speculation with respect to etiology ranging from acoustic shock waves produced by broad beam excimer lasers driving bulk tissue fluid from corneal periphery to corneal center to the adverse effects of the excimer laser plume in blocking incoming laser pulses, little if any actual clinical data exists to support these theories[232,233]. However, our model predicts the occurrence of central islands and provides a physiologic explanation for the etiology of these entities.

> Perhaps no other complication of refractive surgery is as poorly understood as the problem of central islands. Theories on the etiology of central islands are uniformly based upon the idea that such islands occur as a result of aberrancies in the execution of basic shape transfer theory

An analysis of all currently known theories on the etiology of central islands demonstrates that such theories are uniformly based upon the idea that such islands occur as a result of aberrancies in

the execution of basic shape transfer theory. Shape transfer theory proposes that ideal clinical results in laser vision correction are produced when a proposed power shape is accurately reproduced in the corneal tissue. Simply stated, the predominant disease paradigm concerning central islands in the cornea reasons that these islands occur as a product of an improper shape being created on the surface of the corneal stroma due to the presence of a blocking agent (whether that agent be water or smoke) or a central flaw in broad beam laser homogeneity.

> We do not dispute that such deviations in shape transfer are possible, our model predicts that central islands do not typically occur due to the flawed delivery of an ideal shape but rather result because of response characteristics inherent in the cornea. As a result, given the right set of physiological conditions, they are predictable events

Although in theory we do not dispute that such deviations in shape transfer are possible and may occur on occasion, our model predicts that central islands do not typically occur due to the flawed delivery of an ideal shape but rather result because of response characteristics inherent in the cornea. We propose that central islands occur as a result of how the cornea physiologically responds to the event of laser vision correction surgery rather than adhering to what we consider to be simplistic views that the cornea acts like an inert piece of plastic wherein the final shape is created predominantly by the laser beam ablation profile present at the completion of surgery.

Although a discussion of the specific details of how corneal physiology is affected by certain types of excimer lasers or how corneal tissue responds to various physical and pharmacologic factors is beyond the scope of this particular paper, we briefly suggest the following principles.

1. Central islands occur due to surgically induced disruptions in the control of corneal Pif and the normal Pif gradient.
2. Central islands occur in the center of the cornea because the center of the cornea responds differently to surgical stress than the corneal periphery
3. As previously described in this paper, there is a marked gradient in the negativity of Pif when comparing the periphery of the cornea to the corneal center
4. This Pif gradient is believed to produce a directionality of fluid movement in the cornea in which fluid generally moves from periphery to center
5. Surgical events such as laser vision correction cause inflammation and the production of inflammatory cytokines
6. Inflammatory cytokines are well known to affect Pif, typically driving it more negative and causing fluid to be imbibed by the tissue inflamed by such cytokines
7. Excimer laser surgery causes keratocyte injury and apoptosis in the region of photoablation. Such injury is more problematic in broad beam lasers that create more thermal and acoustic injury than modern small scanning beam systems
8. Keratocyte disruption results in alteration of local Pif control. Fluid movement to the central cornea is increased due to central Pif becoming highly negative.
9. This results in a temporary localized accumulation of fluid in the anterior central region of the cornea. If severe, the effect may become permanent, particularly if the patients corneal response is blunted by use of topical steroids
10. This local accumulation of fluid and stromal thickening in the center of the cornea is responsible for the development of central islands

Our model predicted that central islands and erratic clinical outcomes would be more probable with the Alcon LADARVision6000™ protocol than with alternative photoablation procedures

To illustrate the clinical relevance of this proposed mechanism in the creation of central islands, we suggest that analysis of a recent clinical problem concerning the etiology of central islands associated with the Alcon LADARVision6000™ (Alcon Laboratories, Fort Worth, Texas) CustomCornea® protocol may be of value. Specifically, our model predicted that central islands and erratic clinical outcomes would be more probable with this specific protocol than with alternative photoablation procedures. These clinical issues had also been seen in previous versions of that excimer platform.

Unfortunately, this problem with central islands ultimately led to the abandonment of this excimer laser platform (Bell K. Urgent: Device Safety Alert. LADARVision6000™ Excimer Laser. "...immediately discontinue performing CustomCornea® myopia procedures using your LADARVision6000™ Excimer Laser...This action is being taken in response to Alcon's receipt of reports from seven surgical centers citing topographically observed "central islands" in patients following custom laser procedures...". Vice President, Quality Assurance. Alcon Refractive Horizons, Inc. February 21,2007).

> Based upon the conventional belief that central islands occur as a consequence of problems associated with shape transfer, the occurrence of severe central islands with the LADARVision6000™ CustomCornea® protocol and the subsequent permanent recall and abandonment of the platform by the manufacturer of that product is particularly enigmatic

Based upon the conventional belief that central islands occur as a consequence of problems associated with shape transfer, the occurrence of severe central islands with the LADARVision6000™ CustomCornea® protocol and the subsequent permanent recall and abandonment of the platform by the manufacturer of that product is particularly enigmatic. Specifically:

1. Anomalous "bulk fluid" movement is only believed to occur when the cornea is subjected to large acoustic shock waves from broad beam lasers. The LADARVision6000™ device is a small beam system and should be immune from such effects[232,233]. Moreover, this effect was not observed with traditional or non-custom LADARVision® ablation profiles.

2. In the context of a small spot scanning laser system with a Gaussian beam profile, laser beam homogeneity is essentially irrelevant as a possible etiology

3. Laser plume blocking could occur and could cause a central island. However;
   a. Central islands did not occur with other LADARVision® models with relatively comparable repetition rates and vacuum systems
   b. Other small spot scanning laser systems (i.e. Technolas 217) that do not use any plume evacuator have not been reported to produce central islands
   c. Other small spot scanning systems that produce substantially higher laser beam firing rates (i.e.. Allegretto Wave™) and therefore presumably produce a larger problem with plume evacuation are not known to produce central islands

4. A software error or hardware malfunction could cause either an isolated central island or even a protracted series of central islands. However, in the context of the modular design and manufacturing process for excimer lasers, an isolated component failure would generally lead to implementation of a modular

software or hardware solution rather than to the permanent recall or abandonment of the entire product.

Clearly, the shape transfer paradigm appears unhelpful in solving this conundrum. However, the behavior of Alcon in response to this paradox is equally puzzling. The financial investment by Alcon in the acquisition of the LADARVision® system from Summit was substantial. Moreover, the FDA approvals for wavefront guided treatments, obtained at significant cost by Alcon, remain some of the broadest in the industry. The decision to abandon this costly investment, to bear the embarrassment and the cost of ill will generated from deserting its user base, to absorb the cost of the scrapping of sophisticated international manufacturing facilities in combination with the subsequent acquisition (at substantial cost) of an excimer laser device that has a slower tracking system, no cyclotorsional control, absolutely no wavefront guided ablation FDA approvals or a wavefront analyzer with a comparably broad dynamic range is both troubling and baffling.

If one assumes that laser plume effects cause these central islands, historically the typical solution to this issue is the installation of a better vacuum system. Although not insignificant, the overall cost for such a solution is in the order of hundreds of dollars per laser system. Certainly, the cost for the development and installation of a stronger vacuum pump would be insignificant compared to the path chosen by Alcon management.

The possibility that a rare software or hardware problem created these islands could represent a more challenging problem. However, since non-custom photoablations with the LADARVision® systems did not report central islands as a complication, trouble shooting or even reverse engineering the solution for CustomCornea® appears possible. Even if the error in beam delivery is complex it should not be insurmountable. However, even in the circumstance of a complex solution, the cost to repair the system remains extremely small compared to choosing to not repair the device.

We note that a particularly unique aspect of the LADARVision® system is that for each laser pulse delivered by each laser system during a patient procedure, the precise position of the beam and each mirror in the delivery system is recorded in the computer system. In that setting, troubleshooting for a software glitch or a hardware malfunction such as a galvo mirror motor error is a straightforward process. Once the source of the mirror or beam position fault is identified a software or hardware solution can be implemented. Even if one postulates that the software error or hardware failure was severe or occurred across all installed platforms, such an error in one module is not catastrophic. Irrespective of the complexity of a solution to an error in one module of the system, the cost to implement a solution for such a problem remains insignificant compared to other chosen alternatives. Moreover, the field implementation of such a solution is common in the industry and is governed by relatively straightforward FDA protocols.

> Alcon engineering and management were unable to reconcile the lack of congruity between the observation that central islands occurred in CustomCornea® treatments and their inability to find a shape transfer error to explain why these complications occurred. Without a shape transfer fault as the source of the problem the excimer laser system is essentially unfixable. Without a solution, there is no other alternative. The laser platform had to be abandoned

We believe that Alcon engineering and management were unable to reconcile the lack of congruity between the observation that central islands occurred in CustomCornea® treatments and their inability to find a shape transfer error to explain why these complications occurred. In the absence of their ability to clearly identify a shape transfer fault as the source of the problem, the excimer laser system is essentially unfixable. Without a solution, there is no other alternative. The laser platform had to be abandoned.

We believe that our model successfully predicted the failure of the LADARVision6000™ CustomCornea® system. The basis for that prediction is our understanding of the physiologic principles controlling Pif and corneal shape and specifically how these are adversely impacted by the LADARVision® surgical processes.

We (BRW) used the LADARVision® platform in its various device iterations on over 10,000 laser vision correction procedures from 1999 through 2004. Extensive trend analysis of clinical results from this period allowed us to identify what we perceived to be a propensity of this device to produce erratic and less than predictable refractive results under the following circumstances:

1. Surgical delay. Patients that were pharmacologically dilated for an extended period of time (> 45 minutes) prior to actual laser treatment appeared to demonstrate significantly increased risk for visually significant residual refractive error

2. Poor beam quality and inefficient photoablation. When the "volume per shot" or VPS of the laser beam was below 430 mm^3 the risk for visually significant residual refractive error appeared to increase

3. Poor epithelial health. Patients with mild dry eye or preoperative breakdown of the epithelial barrier appeared to be subject to increased risk for visually significant residual refractive error

Based on our model and clinical experience, we postulated the following:

1. Prolonged exposure of the cornea to phenylephrine prior to laser surgery adversely affected the control of Pif in the post LASIK eye. Marinating the eye in phenylephrine preoperatively subsequently induced the synthesis of inflammatory cytokines that affected keratocyte function. This is a G-protein modulated time dependent biological process. When the excimer laser was applied to such a cornea, the adverse effects of the thermal and

acoustic effects of the laser beam on keratocyte behavior were significantly worsened. As a result, control over Pif, fluid dynamics and corneal shape in the healing period was compromised.

2. Poor beam quality and efficiency subjected the cornea to increased levels of thermal loading. Each pulse of the LADARVision® delivers approximately 2.7 mJ of laser energy, irrespective of VPS. Since energy is neither created nor destroyed, thermal effects from excessive sub-threshold UV photons in inefficient laser pulses further potentiate the adverse effects of phenylephrine on corneal fluid dynamics by contributing to the damage to keratocyte function through inside-out transmembrane signaling or the induction of apoptosis.

3. The moderately increased laser firing rate of the 6000 system compared to the 4000 system created more thermal damage to retroablation zone keratocytes.

4. Eyes with poor epithelial health inherently produce a greater amount of baseline inflammatory cytokines as well as demonstrating an increased penetrance for phenylephrine. These effects serve to increase the damage to keratocyte health and the disruption of control of Pif.

In contrast to conventional laser surgery with the LADARVision® system, the CustomCornea® protocol generally involves the dilation of the patient on the day of surgery prior to wavefront measurement and the actual procedure. In that circumstance, the wavefront is also

Based on our experience and understanding of Pif control of the cornea, our model successfully predicted that the LADARVision6000™ CustomCornea® protocol would result in a significant increase in the proportion of patients that would experience unpredictable refractive outcomes, induced hyperopia, induced astigmatism, central islands and an increased incidence of epithelial ingrowth compared to conventional LADARVision® procedures

typically reviewed by the surgeon and processed for treatment. The time required to accomplish this sequence of events frequently causes the patient to be exposed to pharmacologic agents including phenylephrine for an extended period of time prior to surgery. Wait times for patients in the CustomCornea® protocol that approximated one hour in total were apparently not unusual in the clinical setting (Probst LE. Timed Study of Wavefront Evaluation and Treatment with 3 Systems. Scientific Paper. American Society of Cataract and Refractive Surgery. Washington DC. April 2005). Since the effect of phenylephrine is a G-protein mediated process, its effects are time dependent with amplification of effect occurring with increasing time intervals of drug exposure. Multi-dosing of other topical agents such a Proparacaine instilled for astigmatism marking further damage the epithelium, induce the production of additional inflammatory cytokines and increase the penetrance of phenylephrine.

Based on our experience, guided by our understanding of Pif control of the cornea, our model successfully predicted that the LADARVision6000™ CustomCornea® protocol would result in a significant increase in the proportion of patients that would experience unpredictable refractive outcomes, induced hyperopia, induced astigmatism, central islands and an increased incidence of epithelial ingrowth compared to conventional LADARVision® procedures. In addition, our model predicts that these problems associated with the LADARVision® system can largely be resolved by:

1. Eliminating the use of phenylephrine as a dilating agent
2. Consistently providing a high quality beam profile with an efficient photoablation footprint that delivers less thermal injury and UV noise
3. Reducing the instillation of topical agents that increase ocular irritation in the pre and postoperative period

Fortunately, the experiment designed to prove these concepts has already been completed. It is called the Wavelight Allegretto Wave™ Excimer Laser System, (WaveLight Laser Technologie AG, Am Wolfsmantel 5, 91058 Erlangen, Germany). Despite a slower tracking system, no cyclotorsion control and the absence of true wavefront guided ablation protocols, the Allegretto Wave™ (which similar to the LADARVision® system, uses a scanning Gaussian shaped beam) reportedly provides highly predictable refractive endpoints and no central islands

(US FDA clinical trials, data on file with FDA).

Our model of corneal fluid dynamics in the cornea accurately predicted the demise of the LADARVision6000™ excimer laser platform and the occurrence of central islands and erratic outcomes when using CustomCornea®. These events are a product of systematically ignoring the effects of laser vision correction surgery and pharmacologic agents on corneal physiology. Most specifically, they occur because the vision correction industry routinely disregards and generally misunderstands the adverse effects of inflammatory cytokines, outside-in transmembrane signaling and other physiologic forces on the control of Pif, fluid dynamics and corneal shape that occur during and after laser refractive surgery

In summary, through the use of our model of fluid dynamics in the cornea, we were able to accurately predict the demise of the LADARVision6000™ excimer laser platform and the occurrence of central islands when using CustomCornea®. Moreover, these events are a product of systematically ignoring the effects of laser vision correction surgery and pharmacologic agents on corneal physiology. Most specifically, they occur because the vision correction industry routinely disregards and generally misunderstands the adverse effects of inflammatory cytokines, outside-in transmembrane signaling and other physiologic forces on the control of Pif, fluid dynamics and corneal shape that occur during and after laser refractive surgery. Furthermore, these complications are generally not the result of errors in the execution of shape transfer protocols. In fact, in this case troubleshooting from a shape transfer paradigm correctly led to the conclusion that the LADARVision® device exhibited no flaws in its beam delivery algorithms and consequently had no practical solution for complication prevention. Unfortunately, in the absence of an appreciation for how physiologic factors affect corneal physiology and shape and the absence of an identifiable shape transfer error, the LADARVision® platform is an unsafe device. Alcon management correctly rejected its further usage.

Current Flawed Nomenclature

We posit that the current confused and flawed nomenclature can be best characterized as a Tower of Babel debacle. Various observers have attempted to apply names to clinical observations wherein the underlying pathophysiologic principles behind the development of such names are

> We propose to eliminate the current confusing proliferation of names used to describe this single pathophysiologic entity and suggest that the overall constellation of pathology and physiology demonstrated in these 'disorders' be named "keratocyte induced corneal micro-edema" or KME

either flawed or simply erroneous. As a result, in the scientific literature distinctions between such entities as Diffuse Lamellar Keratitis (DLK), Stage 4 DLK, Central Toxic Keratopathy (CTK), Central Flap Necrosis, Flap Necrosis Syndrome and Central Lamellar Keratitis are based on convoluted unscientific theories and a mistaken and frequently magical interpretation of clinical presentations.

We propose to eliminate the current confusing proliferation of names used to describe this single pathophysiologic entity and suggest that the overall constellation of pathology and physiology demonstrated in these 'disorders' be

named "keratocyte induced corneal micro-edema" or KME. Clearly, the fundamental cause of the pathology of these disorders is directly related to the role keratocytes play in controlling Pif and localized tissue edema. Most importantly, the pathology represents a single disease process rather than multiple entities.

> We propose that the incorrect and misleading terminology of Diffuse Lamellar Keratitis (DLK), Stage 4 DLK, Central Toxic Keratopathy (CTK), Central Flap Necrosis, Flap Necrosis Syndrome and Central Lamellar Keratitis be abandoned since this disorder is neither diffuse nor central, it is not a keratitis or a keratopathy, it is not isolated to the LASIK flap and necrosis is not a significant element in the pathophysiologic cascade

In addition, we propose that the incorrect and misleading terminology of Diffuse Lamellar Keratitis (DLK), Stage 4 DLK, Central Toxic Keratopathy (CTK), Central Flap Necrosis, Flap Necrosis Syndrome and Central Lamellar Keratitis be abandoned since this disorder is neither diffuse nor central, it is not a keratitis or a keratopathy, it is not isolated to the LASIK flap and necrosis is not a significant element in the pathophysiologic cascade.

Disease Management

Novel Management Protocols

We suggest that current management protocols fail to provide surgeons or patients with viable treatment options directed towards consistently delivering excellent visual outcomes for DLK, Stage 4 DLK, CTK, CFN, FNS and CLK. This at least partially due to widespread acceptance of incorrect pathophysiologic mechanisms currently promoted as a cause for these disorders.

> The excellent visual results demonstrated by patients in our series demonstrate that protocols directed towards managing disruption in corneal fluid dynamics caused by changes in Pif and tissue tension can dramatically influence clinical outcomes for the better

Based upon the excellent visual results demonstrated by patients in this series, we posit that treatment protocols directed towards managing disruption in corneal fluid dynamics caused by changes in Pif and tissue tension can dramatically influence clinical outcomes for the better.

Our suggestion for management of severe KME (Stage 4 DLK, CTK, CFN, FNS and CLK) involves the following protocols. First, the clinician needs to rule out an infectious keratitis[234]. This can most

> The primary concerns in the management of severe KME include:
>
> 1. Assuring the patient that full visual recovery is highly probable
> 2. Managing the acute edema
> 3. Managing flap macrostriae
> 4. Managing short term and long term residual refractive error, if any

easily be accomplished by noting the pathognomonic presentation combined with the marked disparity between the degree of central corneal whitening observed at the slit lamp compared to the external penlight exam.

Overall, the primary concerns in the management of severe KME include:

1. Assuring the patient that a full visual recovery is highly probable
2. Managing the acute edema
3. Managing flap macrostriae
4. Managing short term and long term residual refractive error, if any

Managing the Edema

Current management protocols suggest lifting and rinsing the interface with sterile balanced salt solution if there are inflammatory cells in the interface. Although well accepted, this treatment has not been subjected to a rigorous clinical trial to demonstrate its efficacy. Based on our model and experience, we believe this course of action to have questionable value. Inflammatory cells in the interface represent a symptom of the disruption in control of Pif. If removing these cells might decrease the production or release of additional inflammatory cytokines then there may be some rationalization for this treatment. We are not aware of any basic science data that would either reject or

> Suppression of the production of inflammatory cytokines through the aggressive short term use of topical steroids is likely the most effective method of preventing and treating mild to moderate KME. However, once the cornea exhibits marked central edema with associated loss of BSCVA, we suggest that steroids be discontinued completely

support such a possibility. However, this should also be balanced with the understanding that this additional surgery will further damage epithelium and disrupt any barrier to fluid influx that epithelial healing may have provided in the hours or days since the original surgery. We also suggest that the common observation that removing the cells results in fewer cells re-accumulating under the flap is more a product of the stroma over time becoming increasingly saturated with fluid resulting in Pif becoming less negative and therefore disposed to drawing less fluid into the interface from the tear film, than evidence that removing cells or some foreign antigen in the interface is beneficial to the final outcome.

Suppression of the production of inflammatory cytokines through the aggressive short term use of topical steroids is likely the most effective method of preventing and treating mild to moderate KME. However, once the cornea exhibits marked central edema with associated loss of BSCVA, we suggest that steroids be discontinued completely. In our opinion, when the cornea is frankly edematous, their adverse effects far out weight the clinical benefit of topical steroids. Oral steroids may have some efficacy but their use should be weighed against potential associated systemic side effects or complications. In general, we are in agreement with Hainline et al.[24] and Sonmez and Maloney[22] in advising against the use of steroids of any kind in severe KME.

> In more severe cases of KME, particularly if the cornea is exhibiting focal central edema with loss of BSCVA, treatment with topical hyperosmotics is very helpful. We believe that treatment with Muro drops can prevent a case from progressing to Stage 4. In severe KME with gross edema we suggest an aggressive "kick start" with topical Glycerol to see how much effect can be anticipated from hyperosmotic treatment followed by a maintenance dose of Muro drops and ointment to optimize vision

In more severe cases of KME, and particularly if the cornea is exhibiting focal central edema with loss of BSCVA, treatment with topical hyperosmotics is very helpful. If the patient has both eyes affected then this treatment is generally more readily accepted than if the patient retains good vision in the other unaffected eye. In either event, if the patient wishes to improve vision over the short term the treatment we suggest involves an aggressive "kick start" to see how much effect can be anticipated from hyperosmotic treatment followed by a maintenance dose to optimize vision.

The initial treatment involves applying one drop of Proparacaine in the affected eye followed by Glycerol PF 99.5% 1 gtt q 10 mins X 3 (Leiter's Pharmacy, San Jose, CA, USA) while the patient is in the exam lane. If BSCVA improves by 2 or 3 lines on the Snellen chart with this regimen, there is good evidence that the patient will experience better visual quality with a maintenance course of topical hyperosmotics. We recommend use of Muro 128® q1-2 hr and Muro 128® ung qhs. The patient should be alerted that these drops burn on instillation. This burning does has the effect of limiting compliance. However, patients with severe bilateral involvement may find that the benefit warrants the discomfort. Although we believe that treatment with hyperosmotics can rapidly reverse the edema causing the clinical manifestation of KME and accelerate recovery, at this time we cannot determine if such a treatment actually improves long term outcomes. The patient should also be placed into a soft contact lens to enhance corneal dehydration and correct refractive error.

We also suggest the patient use Acular® PF QID and oral Vitamin C 1000 mg/day. Tanaka et al.[235] demonstrated a marked

We also suggest the patient use a non-preserved topical NSAID such as Acular® PF QID and oral Vitamin C 1000 mg/day. Vitamin C has demonstrated a marked therapeutic effect in the reversal of negative Pif and early edema generation in burns

therapeutic effect of high dose Vitamin C in the reversal of negative Pif and early edema generation in burns. In the Vitamin C groups, there was a marked attenuation of the negative Pif created by thermal burn and a marked reduction in interstitial tissue edema. Williams and Paterson[236] also demonstrated that elevated ascorbate concentration in the aqueous humor had a marked inhibitory effect on lipoxygenase synthesis within the cornea. Lipoxygenase metabolites have previously been shown to directly cause corneal edema[42]. Kasetsuwan et al.[237] also found that topical ascorbic acid application following PRK in rabbit eyes resulted in significantly reduced acute inflammation and oxygen radical tissue damage, both factors being known causes of corneal edema. Stojanovic et al.[238] demonstrated that high dose vitamin C taken preoperatively markedly decreased the incidence of postoperative corneal haze in PRK. Although there is no direct link between corneal haze in PRK and corneal edema, be believe that our model suggests that they are causally related.

Managing Macro and Microfolds in the LASIK Flap

In our experience, the primary (if not only) mechanism for loss of BSCVA in KME is persistent microstriae in the visual axis. As a result, we recommend that they be aggressively managed. However, the key to managing residual microstriae that adversely affect BSCVA is to allow the control of Pif to become normalized in the cornea prior to surgical

> The primary (if not only) mechanism for loss of BSCVA in KME is persistent microstriae in the visual axis. We recommend that they be aggressively managed. The key to managing residual microstriae that adversely affect BSCVA is to allow the control of Pif to become normalized in the cornea prior to surgical intervention

intervention. In practice, we suggest that this involves allowing keratocyte population levels and cell function to recover towards baseline. In our experience this appears to be accomplished by approximately 6 to 8 weeks from the onset of severe KME. If treatment is delayed much longer than this, the microstriae may become more difficult to completely remove. Certainly, if the patient is correctable to a sharp refractive endpoint and the microstriae do not create unwanted images or distortions, they are simply best left alone.

> Management of residual visually significant microstriae involves complete removal of the epithelium overlying the striae as a first step. Once the central epithelium is completely removed the entire flap can be lifted, refloated and repositioned on the bed. Once repositioned the flap should be aggressively stretched with dry merocels for 2 or even 3 minutes. There is no need for heat, hyposmotic agents or sutures. One drop of Glycerol PF should be instilled on the center of the flap followed by a soft bandage contact lens for at least 24 hours or more, depending on the size of the central epithelial defect

Management of residual visually significant microstriae involves complete removal of the epithelium overlying the striae as a first step. This can be easily done using a #64 Beaver® blade under the surgical microscope found on the excimer laser. The epithelium should be removed in our experience because the hemidesmosomes and tight junctions in the central epithelium seem defective and allow fluid to more easily move into the adjacent stroma than normal. They may also cause some tethering of flap macrofolds allowing them to resist manual stretching maneuvers. Once the central epithelium is completely removed the entire flap can be lifted, refloated and repositioned on the bed. The bed and undersurface of the flap can be cleaned or even scraped if desired, however we fail to see any therapeutic value in such tissue manipulation. There is nothing present in the interface or on the surface of the flap or residual bed that needs to be physically removed. Once repositioned the flap should be aggressively stretched with dry merocels for 2 or even 3 minutes. There is no need for heat or hyposmotic agents. Sutures are not needed and we do not recommend that they be used. When complete, the flap edge should be well apposed to the flap sidewall with a normally sized gutter. The flap may not be completely smooth in appearance.

> The patient should be discharged on a slowly tapered regimen of topical corticosteroids q2h, broad spectrum topical antibiotics QID, Muro 128® q1-2h, Muro 128® ung qhs, a topical preservative-free NSAID QID and Vitamin C 1000 mg PO qd.

One drop of Glycerol PF should be instilled on the center of the flap followed by a soft bandage contact lens for at least 24 hours or more, depending on the size of the central epithelial defect. We recommend use of a bandage contact lens until the epithelial defect is fully healed. The patient should be discharged on topical corticosteroids q2h, broad spectrum topical antibiotics QID, Muro 128® q1-2h, Muro 128® ung qhs, a topical preservative-free NSAID QID and Vitamin C 1000 mg PO qd. This regimen should be gradually tapered over time based on the best clinical judgment of the clinician involved in the patients care.

Management of Residual Refractive Error

In our opinion, the presence of a residual refractive error in KME once corneal healing is completed is an uncommon outcome. However, larger long term studies will be required to confirm the true incidence of this problem. If microstriae persist, it may be reasonable to consider a second lift and stretch procedure combined with central epithelial removal.

> If residual refractive error persists in the absence of clinically significant microstriae, we suggest excimer laser retreatment. Wavefront or topographically driven photoablation profiles may be superior to traditional non-custom treatments. However, unless the original photoablation was abnormal, a custom approach is not truly needed as no clinically relevant flap or bed necrosis occurs in KME.

However, if residual refractive error persists in the absence of clinically significant microstriae, we suggest excimer laser retreatment. Wavefront or topographically driven photoablation profiles may be superior to traditional non-custom treatments. However, all of the patients in our series were managed with non-custom treatments. Unless the original photoablation was abnormal, a custom approach is not truly needed as no clinically relevant flap or bed necrosis occurs in KME.

Regardless of which treatment approach is taken, it is important to allow the cornea to stabilize physiologically and obtain a stable refraction before retreatment with

> We suggest waiting at least 3 to 6 months before retreatment can be considered with the final decision based upon the stability of serial refractions, serial wavefront measurement and serial corneal tomography. Pretreatment with steroids or other atypical management rituals are not suggested or needed as there is no greater risk for recurrence of KME than in any routine LASIK retreatment. The patient should be expected to obtain an excellent visual outcome

the laser. We suggest waiting at least 3 to 6 months before retreatment can be considered. The final decision should be based upon the stability of serial refractions, serial wavefront measurement and serial corneal tomography. During enhancement, the eye can be treatment normally. Pretreatment with steroids or other atypical management rituals are not suggested or needed as there is no greater risk for recurrence of KME than in any routine LASIK retreatment. The patient should be expected to obtain an excellent visual outcome.

Prevention of KME

A comprehensive discussion of how to best prevent KME is not possible in this paper. Our model predicts that KME is best prevented by:

1. Reducing ocular inflammation pre and postoperatively

2. Reducing the adverse effects of intraoperative tissue shear or crush injury as well as damage from thermal and acoustic shock

3. Reducing the antigenic or inflammatory stimulus presented to the ocular system during surgery

4. Reducing the release and production of inflammatory cytokines in the eye during and after laser refractive surgery

5. Reducing factors that cause outside-in transmembrane signaling during the procedure

6. Eliminating or reducing the use of pharmacologic agents that act as cytokine analogs or otherwise adversely affect Pif through cytokine production or by directly modulating keratocyte behavior by binding to appropriate receptors on keratocyte cell membranes

7. Eliminating or reducing the use of agents such as organic solvents, BAK and topical anesthetics that intercalate within cell membranes, alter the dynamics of the phospholipid bilayer and induce conformational changes in transmembrane proteins or otherwise affect calcium or cAMP balance in the cell

Certainly, performing surgery in a manner that reduces the introduction of antigens and toxins or suppresses the induction of inflammatory cytokines in the eye is an important first step. Minimizing lipopolysaccharide loading, eliminating contaminants from drapes or surgical gloves, reducing aerosolized antigens or the introduction of tissue toxins during and after surgery are important factors. In addition, minimizing injury to the epithelium and corneal stroma are also key elements in KME prevention. Reducing pulse energy and optimizing pulse separation when using femtosecond lasers will prevent KME. Similarly, reducing the thermal and acoustic footprint of the excimer laser will also improve keratocyte health and measurably reduce the incidence of KME. In addition, attempting to modulate the inflammatory response of the host through the use topical[1,2,4-12,15,239,240], oral[23,47] or intrastromal[59] steroids as well as topical NSAID's[241,242] is believed to be of significant benefit.

Possible Experimental Treatments

Our model suggests several useful avenues for research on the treatment and prevention of KME.

1. Blocking the synthesis of inflammatory cytokines
2. Blocking the effects of inflammatory cytokines
3. Reversing changes induced in Pif
4. Reducing the mechanical effect of IOP on tissue compliance and reducing backflow of fluid into the stroma by reducing IOP
5. Using endothelial pump enhancers

Blocking the upregulation of inflammatory cytokine cascades can be accomplished through the use of corticosteroids, NSAID's and Histamine receptor-1 (H_1) antagonists. Steroids, whether given topically, orally or intrastromally are well known to block the activity and reduce the synthesis of

> Our model suggests several useful avenues for research on the treatment and prevention of KME.
>
> - Blocking the synthesis of inflammatory cytokines
> - Blocking the effects of inflammatory cytokines
> - Reversing changes induced in Pif
> - Reducing the mechanical effect of IOP on tissue compliance and reducing backflow of fluid into the stroma by reducing IOP
> - Using endothelial pump enhancers

inflammatory cytokines by inhibiting ICAM-1 expression, IL-1, TNF and IL-8 secretion and otherwise inhibiting the upregulation of the inflammatory cascade that can lead to KME. In addition, steroids generally inhibit the infiltration and accumulation of neutrophils in tissue.

Inhibitors of prostaglandin synthesis such as cyclooxygenase inhibitors may have beneficial effects in preventing trauma-induced edema in refractive surgery, if they are dosed preoperatively and postoperatively. Studies by Ruberti et al. [163] and Karon and Klyce[164] both clearly demonstrate the edema reducing properties of the COX inhibitor diclofenac in a rabbit model. Similarly, Sandoval et al.[241] found that topical ketorolac tromethamine 0.5% showed a tendency towards a lower rate of DLK and reduced severity of the disease in a rabbit model when DLK was induced by LPS and Palmolive soap.

> Blocking the upregulation of inflammatory cytokine cascades can be accomplished through the use of corticosteroids, NSAID's and Histamine receptor-1 (H₁) antagonists

Boorstein et al.[45] demonstrated that non-sedating H_1 antagonists reduced the incidence of DLK in atopic individuals undergoing LASIK surgery. As a mechanism for this clinical effect, they postulated that oral or topical non-sedating H_1 antagonists may inhibit ICAM-1 expression as well as suppress secretion of IL1, TNF, and IL-8 in eyes undergoing laser refractive surgery. Holzer et al.[243] found that the mast cell stabilizer nedocromil sodium ophthalmic solution 2% (Alocril®, Allergan, Irvine, CA, USA) reduced the severity of DLK in a rabbit model where DLK was induced by LPS and suggested it may have clinical therapeutic value.

Specific blockers for endotoxins[244] and PAF[20] have been previously considered as useful agents in preventing or treating DLK. Our model similarly predicts that they will be highly useful in the management of KME. Morck et al.[244] suggested that the endotoxin blocking activity of polymyxin B may be useful in reducing the incidence of DLK in a rabbit model where DLK was induced by purified LPS isolated from *Burkholderia cepacia*. Esquenazi et al.[20] reported on the peribulbar injection and topical use of the PAF receptor antagonist LAU 0901 in preventing the induction of DLK by Pseudomonas LPS in rabbit eyes. They found a marked decrease in incidence and

severity of DLK as well as a reduction in keratocyte apoptosis, a reduced infiltration by inflammatory cells into the interface and a decrease in myofibroblast formation. Those results, further commented upon by Bazan[42] in a review of lipid signaling in the cornea, suggest that inhibition of PAF synthesis or the blockage of PAF receptors may be a potent tool in the prevention of KME.

Reversing changes in Pif induced by cytokines, chemokines and outside-in transmembrane signaling may be accomplished by using topical hyperosmotics or other agents known to alter Pif. Our results using topical glycerol indicate that in select cases, it can

> Specific blockers for endotoxins and PAF such as polymyxin B and the PAF receptor antagonist LAU 0901 have been previously considered as useful agents in preventing or treating DLK. Our model similarly predicts that they will be highly useful in the management of KME

be very effective in reducing edema in the cornea in KME and accelerating recovery of vision.

Berg et al.[139] demonstrated that lantanoprost (a $PGF_{2\alpha}$ analog) reversed in a dose dependent fashion the increased negativity of Pif created by the anaphylactic response to dextran in rats. Stuhr et al.[245] used a human dermal fibroblast cell aggregate model to demonstrate that $PGF_{2\alpha}$ had the effect of making Pif more positive. We believe that these studies suggest a novel therapeutic use for lantanoprost (Xalatan®, Pfizer Inc., New York, NY, USA) in the control of tissue edema in the cornea in the perioperative and postoperative phase of laser refractive surgery as well as in the management of KME.

Other agents such as α-trinositol or the BB isoform of PDGF that have demonstrated an ability to cause contraction of 3-dimensional collagen

> Reversing changes in Pif induced by cytokines, chemokines and outside-in transmembrane signaling may be accomplished by using topical hyperosmotics or other agents known to alter Pif such as lantanoprost, α-trinositol, the BB isoform of PDGF and Vitamin C

matrices populated by fibroblasts may be potentially highly useful in controlling corneal fluid dynamics during and after laser refractive surgery. The drug α-Trinositol, an anti-inflammatory agent that apparently works by altering intracellular calcium channels[146], has been demonstrated to either abolish or strongly attenuate the fall in Pif and edema formation in burns[147], dextran anaphylaxis[148], blockade of β1–integrins[149] and local frostbite injury[150]. It has also been demonstrated to have similar effects in the trachea in experimental asthma[151] and neurogenic inflammation[152].

Another potent stimulator of collagen gel contraction in vitro and in vivo is the BB isoform of platelet-derived growth factor (PDGF-BB). Rodt et al.[153] demonstrated that after dextran anaphylaxis had induced

lowering of Pif in the rat paw, PDGF-BB injected subdermally brought the Pif completely back to normal. Linden et al.[154] found that the ability of PDGF-BB to counteract the tendency towards edema was accomplished by the stimulation of αVβ3-integrins. The specific action of PDGF-BB have been further studied by Heuchel et al.[155] in mutant mice, wherein their group was able to demonstrate that the PDGF β-receptor operates via the phophatidylinositol 3'-kinase pathway and has a role in control of tissue fluid homeostatis in vivo by a direct effect on interstitial fluid pressure. Moreover, Ahlen et al.[146] demonstrated that cell-collagen interaction both in vivo and in vitro depend on phosphatidylinositol 3'-kinase, and that this dependence can be bypassed by a drug eliciting intracellular Ca^{2+} mobilization.

The observation that both α-Trinositol and PDGF-BB demonstrate the ability to reverse the lowering of Pif suggests that they may be useful in the control of KME and the modulation of corneal fluidics. Such agents may also demonstrate an ability to produce a more predictable refractive endpoint when used in routine refractive surgery. Vitamin C given topically or orally may also be an adjunct to managing KME and enhancing the accuracy of refractive surgery as it has been shown to diminish tissue edema[235-237]. Basic science research with these and similar compounds may provide additional direction.

Elevation of IOP has been shown to be a significant factor in the development of IFS[204-208,246]. It is also possible that

Reduction of IOP using anti-glaucomatous medication will be beneficial in reducing or controlling the clinical effects of KME by reducing "back pressure" and the retrograde flow of fluid from the anterior chamber into the corneal stroma. Reducing IOP may also reduce tension placed on the corneal stroma thereby improving tissue compliance and reducing the opportunity for compartment black and severe KME

reduction of IOP through the use of anti-glaucomatous medication will be beneficial in reducing or controlling the clinical effects of KME by reducing "back pressure" and the retrograde flow of fluid from the anterior chamber into the corneal stroma. Reducing IOP may also reduce tension placed on the corneal stroma thereby improving tissue compliance and reducing the opportunity for compartment black and severe KME to occur.

Leu and Hersh[247] describe the use of phototherapeutic keratectomy (PTK) for treatment of recalcitrant DLK. Our model suggests that this methodology may not be ideal as a treatment modality for DLK and we do not advocate its use for KME.

Improving the Predictability of Refractive Results

Model Defines a New Pathway Forwards

We suggest that our unifying model provides a framework upon which to build an improved understanding of corneal physiology and the impact of laser vision correction procedures on corneal homeostasis. Future clinical and basic science research will likely improve our understanding of the role of keratocytes in

> Advancements and insight into the precise relationship between Pif, tissue tension and tissue compliance will markedly improve our ability to deliver more accurate refractive endpoints as well as delivering many fewer surgical complications

maintaining normal tissue hydration. In addition, as we begin to better understand the effects of inflammatory cytokines, thermal and mechanical injury, acoustic shock waves, femtosecond lasers, pharmacologic agents, aerosolized antigens or tissue irritants, pharmacologic agents, organic solvents and other tissue toxins on keratocyte function and control of Pif, our clinical outcomes and incidence of surgical complications will markedly improve. We also believe that advancements and insight into the precise relationship between Pif, tissue tension and tissue compliance will markedly improve our ability to deliver more

accurate refractive endpoints as well as delivering many fewer surgical complications.

Model Applies to Predictability of Refractive Outcomes

Although not described in any detail in this paper, our model also predicts that the impact of biologically mediated biomechanical shape alteration in the cornea is also a significant factor contributing to the predictability, or lack thereof, of the refractive endpoint of laser

> Our model predicts that the impact of biologically mediated biomechanical shape alteration in the cornea is a significant factor contributing to the predictability, or lack thereof, of the refractive endpoint of laser vision correction surgery

vision correction surgery. Currently, we are capable of accurately measuring refractive errors preoperatively to approximately ± 0.01 D and sophisticated excimer lasers remove tissue with submicron precision. However, our expected refractive results frequently differ from predicted by 1 diopter or more thereby representing a 100X error rate.

Our model suggests that factors that create excessive hydration of the cornea will tend to induce hyperopic shifts, while those that induce less hydration than expected will trend towards producing

> Our model predicts that myopic corrections will cause excessive central flattening and excessive peripheral steepening and a tendency to produce hyperopic overshoots and increased amounts of positive spherical aberration. Conversely, our model predicts that hyperopic corrections will cause excessive central steepening and excessive peripheral flattening and will therefore tend to deliver myopic overshoots

myopic shifts. In addition, the biomechanical components of our model predict that myopic corrections will cause excessive central flattening and excessive peripheral steepening with the resulting tendency to produce hyperopic overshoots and increased amounts of positive spherical aberration. Conversely, our model predicts that hyperopic corrections will cause excessive central steepening and excessive peripheral flattening and will therefore tend to deliver myopic overshoots. We also believe that the biomechanical effects defined by our model also predict that the clinical impact of hyperopic corrections in producing excessive central steepening and peripheral flattening will be substantially higher on a per diopter basis than that observed in myopia. We believe that these predictions are consistent with clinical results produced by LASIK laser algorithms as well as accepted concepts based upon eye modeling[184-187,248-250].

Based on our model, we posit that transient and permanent overcorrection and undercorrection of refractive error in "uncomplicated" PRK, LASIK or LASEK are, in part, related to biomechanical

> We believe that a better understanding of those factors controlling Pif, tissue tension, tissue volume and tissue compliance must be attained if we anticipate increasing accuracy of outcomes in laser refractive surgery significantly over current levels

forces described in our model. Reducing tissue thermal loading by moving from broad beam to small spot scanning excimer laser systems, decreasing tissue injury from excimer or femtosecond laser induced shockwaves, minimizing the introduction of antigens, eliminating the use of pharmacologic agents such as alpha-agonists that directly disrupt control of Pif, reducing the introduction of tissue toxins and the concomitant use of more specific or powerful anti-inflammatory pharmacologic agents have markedly improved clinical results in recent years.

However, we believe that a better understanding of those factors controlling Pif, tissue tension, tissue volume and tissue compliance must be attained if we anticipate increasing accuracy of outcomes in laser refractive surgery significantly over current levels. Certainly, we suggest that such an understanding is essential in order to gain appropriate insight into the etiology and management of KME.

Conclusions

We propose a pathophysiologic model for DLK, Stage 4 DLK, CTK, CFN, FNS and CLK that postulates that they are all caused by the loss of control of Pif. Inflammatory cytokines, pharmacologic agents and injury to keratocytes disrupt stromal regulated homeostasis of Pif by disrupting ECM-keratocyte interactions. The disruption of these interactions leads to matrix relaxation and drives Pif more negative. The force generated draws fluid and inflammatory cells into the interface and stroma and leads to the formation of tissue edema. Alteration in Pif creates a cascade of physiologic processes that reorder corneal fluid dynamics, alter tissue compliance, redistribute tissue tension across the cornea and initiate reversible biologically mediated mechanical processes. Reversible biomechanical forces cause marked shifts in refraction, transient opacification of central flap stroma and tissue macrofolds.

We assert that control of Pif in the cornea is disrupted during laser refractive surgery by two different mechanisms or cascades converging on a final common pathway. The Inflammatory Arm or 'DLK arm' of the pathophysiologic cascade includes the effects of inflammatory cytokines and mediators including, but not limited to, various forms of IL, TNF, PAF, LPS and prostaglandin derivatives that bind to receptors on keratocyte membranes and create a physiologic effect through the modulation of G-protein directed cell functions. The Non-inflammatory Arm or 'CTK arm' of the pathophysiologic cascade includes factors that directly affect keratocyte function and viability. These factors include the modulation of keratocyte function resulting from mechanical ECM-cell "outside-in" β-integrin mediated transmembrane signaling processes, G-protein modulated effects and the disruption of cell membrane function and architecture. In clinical practice, these latter factors include both mechanical and thermal laser-tissue interactions, the effects of pharmacologic agents that bind to G-protein coupled receptors on keratocyte cell membranes and effects of toxic agents such as organic solvents, topical anesthetics and preservatives such as benzalkonium chloride (BAK) that interfere with cell membrane homeostasis. We assert that rather than being independent, these mechanisms conjointly contribute to the development of this disorder.

Inflammatory cells present in the tear film are drawn into the interface due to decreased Pif and bulk fluid flow toward the corneal center, rather than by the directional force of chemotaxis. They are a symptom rather than a cause of the disorder. DLK is largely an inflammatory process that affects keratocyte function and can frequently be non-specifically suppressed with steroids. CTK or the non-inflammatory arm occurs primarily as a result of keratocyte injury or, occurs in a setting where the constellation of inflammatory cytokines and mediators produced does not create a significant recruitment of acute inflammatory cells into the tear film. We suggest and the literature demonstrates significant overlap in clinical presentations between these entities. However, for both disorders, the primary central pathology is caused by tissue edema and is not a consequence of collagen or ECM matrix necrosis. As a result, aside from the management of residual flap macrofolds produced by frank tissue edema, both disorders are essentially self-limited. The consistent localization of edema in the center of the cornea occurs as a consequence of a compartment syndrome caused by the anisotropic movement of fluid in the cornea, a mismatch in compliance between the flap and underlying residual stromal bed and fluid trapping in the flap.

The induced hyperopic shift occurs as a consequence of reversible biomechanical forces that result from localized changes in tissue compliance and Pif. These forces are reversible as the cornea recovers from the injury and the effects of inflammatory cytokines. Localized tissue edema creates transient central corneal flattening, mid-peripheral steepening and peripheral flattening combined with compression and thinning of the central residual stromal bed.

In addition, neither Stage 4 DLK, CTK, CFN, FNS or CLK measurably respond to topical steroid treatment as the therapy is either 'too little too late' or has insignificant impact on suppressing corneal edema once the compliance of the anterior interstitial compartment is saturated. Sporadic and late onset cases of interface inflammation are caused by the same mechanism primarily as a consequence of epithelial defects and conjunctival trauma, release of inflammatory cytokines and mediators and uptake of fluid and inflammatory cells from the tear film through a break in the epithelial barrier. The accumulation of acute inflammatory cells in the interface in late onset cases is likely due to the interface representing a path of least resistance as well as fibronectin deposited there acting as a guide and "bread crumb trail" that facilitates inflammatory cell motility and migration.

Appropriate therapy is directed at managing corneal edema acutely and limiting loss of BSCVA through timely management of flap macrostriae. Although there is a predilection for patients with atopy or other pre-existing states of ocular inflammation, in general, this disorder is proportional to the overall corneal and ocular injury as well as the bolus of exogenous antigens that occurs at the time of the initial corneal surgery. As a consequence, it rarely reoccurs with subsequent enhancement procedures. In femtosecond laser procedures; employing strategies that reduce keratocyte injury and inflammatory cytokine production can modulate the intensity of this pathophysiologic cascade. These strategies include reducing raster beam energies and optimizing laser pulse separation, employing continuous plane photodisruption techniques, avoiding use of α-agonists and BAK and the judicious use of NSAIDs and topical steroids.

We further propose to eliminate the current confusing proliferation of names used to describe this single pathophysiologic entity and suggest that the overall constellation of pathology and physiology demonstrated in these 'disorders' be named "keratocyte induced corneal micro-edema" or KME. In addition, we propose that the incorrect and misleading terminology of Diffuse Lamellar Keratitis (DLK), Stage 4 DLK, Central Toxic Keratopathy (CTK), Central Flap Necrosis, Flap Necrosis Syndrome and Central Lamellar Keratitis be abandoned since this disorder is neither diffuse nor central, it is not a keratitis or a keratopathy, is not localized to the LASIK flap and necrosis is not a significant element in the pathophysiologic cascade.

We further posit that a better understanding of those factors controlling Pif, tissue tension, tissue volume and tissue compliance must be attained if we anticipate increasing accuracy of outcomes in laser refractive surgery or reduce the incidence of complications significantly over current levels.

Case Reports

Case 1

JM is 34 year-old Caucasian female whose preoperative manifest refraction was OD – 9.50 and OS -7.75 providing 20/15 OU. Humphrey® Atlas™ corneal topography (Carl Zeiss Meditec, Inc., Dublin, CA, USA) was normal and pachymetry was OD 520 and OS 503 microns.

Bilateral LASIK was performed August 6, 2002. Both LASIK flaps were created in an uncomplicated manner using the Intralase™ Femtosecond laser with an intended flap thickness of 100 microns and a nasal hinge, a raster energy of 2.6 microjoules and a 10 micron spot and line separation with a side cut energy of 5.2 microjoules. An uncomplicated excimer treatment was performed using a LADARVision® excimer laser with an optical zone of 6.5 mm OU with an intended ablation depth of 137 microns OD and 115 microns OS. Postoperatively the patient was treated with Ciloxan® QID (Alcon, Fort Worth, TX, USA), EconoPred® q1hr (Alcon, Fort Worth, TX, USA) and Acular® PF QID OU (Allergan, Irvine, CA, USA).

On POD #1 the UCVA was 20/40 OU with 2+ interface inflammation sparing the visual axis in both eyes. EconoPred® was continued q1hr and Tobradex® (Alcon, Fort Worth, TX, USA) ointment added qhs. On POD #5 the patient presented with features consistent with bilateral Stage 4 DLK with diffuse interface inflammation, bilateral central whitening of the interface, bilateral central micro and macrostriae with UCVA of 20/60 OD and 20/400 OS. Conjunctiva and anterior chamber were normal OU. It was noted that under slit lamp illumination the central interface appeared highly opaque while under direct visualization the central cornea appeared virtually clear.

The OS was immediately lifted with a Hersh spatula and irrigated with 1 cc of sterile balanced salt solution (BSS) (Alcon, Fort Worth, TX, USA) using a Dishler LASIK cannula. At time of surgery it was observed under the surgical microscope that the stromal bed was normal with no evidence for necrosis, tissue loss or opacification. In addition, the flap did not demonstrate any evidence for thinning or opacity. There was no exudates, debris or any abnormal material in the interface. Slit lamp examination performed upon completion of the lift and irrigation procedure demonstrated absence of inflammatory cells in the interface, however the central dense opacification was unchanged from its preoperative condition. The central corneal epithelium was loose so a bandage soft contact lens was inserted. No surgical intervention was performed in the OD. The patient was discharged on Ciloxan® QID, EconoPred® q1hr, Tobradex® qhs and Theratears® prn OU (Advanced Vision Research, Woburn, MA, USA).

On POD #7 (2 days after lift and rinse OS) UCVA was 20/60 OD and 20/200 OS. The OD demonstrated resolving interface inflammation with central corneal opacification and macrostriae. The OS demonstrated negligible inflammatory cells in the interface, a large central / paracentral opacification, central flap macrostriae while epithelial bullae and microcysts were noted overlying the area of central opacification. Intraocular pressure by tonopen was 12 mm Hg OU and there was no interface fluid present in either eye. The Ciloxan®, EconoPred® and Tobradex® were completely discontinued OU. Muro 128® 128 drops (Bausch and Lomb, Rochester, NY, USA) were initiated q2hr OU combined with Muro 128® ointment qhs OU (Bausch and Lomb, Rochester, NY, USA).

At POD #10 while using only Muro 128® drops and ointment the UCVA was 20/40 OD correcting to 20/30 with +0.50 + 0.75 X 165. UCVA OS was 20/200 with no improvement with refraction or pinhole. No change was made to the hyperosmotic treatment regimen.

At POD #13 UCVA OD was 20/40 improving to 20/30 using plano +0.75 X 170. UCVA OS was 20/60 improving to 20/40 using +0.75 +0.75 X 15. Slit lamp finding included diminishing central corneal opacification and resolving macrostriae. No change was made to the hyperosmotic treatment regimen.

At POD #16 UCVA both eyes began to demonstrate a significant hyperopic shift. OD was 20/40+ improving to 20/20- using +2.00 + 1.00 X 175. UCVA OS was 20/50- improving to 20//40- using +1.50 + 1.75 X 65. No change was made to the hyperosmotic treatment regimen.

On POD #19 on the OS the central epithelium was removed using a #64 Beaver blade® (BD, Franklin Lakes, NJ, USA) and the flap was lifted and stretched with dry merocels® (Medtronics, Jacksonville, FL, USA). A bandage contact lens was inserted and the patient continued on Muro 128® gtts q2hr and Muro 128® ung qhs.

At 4 weeks the hyperopic shift in the OD was markedly decreased and the macrostriae were resolving. UCVA OD was 20/40+ improving to 20/15- using +1.00 +1.25 X 175. At 5 weeks the hyperopic shift in the OD was further reduced. Manifest refraction OD was now +0.25 +1.25 X 155 yielding BCVA of 20/20-. However, the manifest refraction in the OS at 2 weeks following the flap lift and stretching procedure (5 weeks from the initial LASIK) demonstrated marked hyperopic shift with UCVA of 20/40 improving to 20/30+ using +3.00 +1.50 X 30 although the flap macrostriae were rapidly improving and the central opacification of the cornea was resolving. No change was made to the hyperosmotic treatment regimen.

At 6 weeks post op (3 weeks from the OS flap lift and stretch) the hyperopic shift was markedly improved in both eyes. UCVA OD was 20/20- improving to 20/20+ using -0.75 +1.00 X 164. UCVA OS remained at 20/40 but improved to 20/25+ using +1.75 +1.50 X 35. Of note was that the hyperopic shift was reduced by fully 1.0 diopter in the OD and 1.25 diopters OS within 7 days and BCVA improved dramatically during this same period using hyperosmotic treatment alone.

At 8 weeks post op (5 weeks from the OS flap lift and stretch) the hyperopic shift continued to resolve dramatically with a concurrent improvement in BCVA in both eyes. As well, the striae in the OD were nearly gone while in the OS the macrostriae were now only faint macrostriae and the sub-interface opacification was minimal. UCVA OD was now 20/20- improving to 20/15 using -1.00 +1.00 X 160. The OS demonstrated a decrease of +1.50 diopters of hyperopia over the preceding 2-week period of time with UCVA OS of 20/30 improving to 20/15- with +0.25 +1.50 X 27. At this examination the Muro 128® regimen was discontinued on the OD and for the OS tapered to Muro 128® gtts QID and Muro 128® ung qhs.

At 12 weeks post op the clinical outcome was demonstrating increasing refractive stability, further reduction in the hyperopic shift and a normal postoperative appearance at the slit lamp with only trace sub-interface haze and macrostriae OS. UCVA OD was 20/20+ improving to 20/15 with Plano + 0.50 X 165. UCVA OS was 20/25 improving to 20/15 using -0.50 +1.25 X 22. All hyperosmotic agents were discontinued.

At 16 weeks UCVA OD was 20/20+ improving to 20/15 using Plano +0.50 X 170. UCVA OS was 20/20- improving to 20/15 using -0.75 +1.00 X 30. There were now only trace macrostriae visible in the OS.

At 6 months the patient underwent an uncomplicated enhancement surgery in the OS using a VISX™ Star S4 excimer laser employing a standard photoablation profile for a sphero-cylindrical correction of -1.00 +1.00 X 25. At one-day post enhancement UCVA OD was 20/20. Follow up at 4 months post enhancement demonstrated UCVA of 20/20 OU with a manifest refraction of Plano OS and with no complaints of glare or halo effects or any other vision complaints. The patient was discharged from our clinic.

Case 2

SL is a 32 year-old Caucasian male whose preoperative manifest refraction was OD -0.75 + 1.00 X 7 and OS Plano +0.25 X 4 demonstrating 20/15 OU. Ultrasound pachymetry was OD 560 OD and OS 567 microns. CustomVue™ WaveScan (Advanced Medical Optics, Santa Ana, CA, USA) testing and Orbscan® II topography (Bausch and Lomb, Rochester, NY, USA) were normal OU.

LASIK was performed on the OD on December 11, 2003. The LASIK flap was created in an uncomplicated manner using an Intralase™ Femtosecond laser settings with an intended flap thickness of 100 microns and a nasal hinge, a raster energy of 2.5 microjoules on a 10 micron spot and line separation with a side cut energy of 5.0 microjoules. An uncomplicated excimer treatment was performed using a VISX™ Star S4 using a standard 5.0 X 6.5 mm optical zone with an intended ablation depth of 9 microns. Postoperatively the patient was treated with Ocuflox® (Allergan Inc., Irvine, CA,

USA), Pred Forte® (Allergan Inc, Irvine, CA, USA) and Acular® PF QID.

On POD #1 the UCVA was 20/25 with 2+ interface inflammation sparing the central 2 mm visual axis. On POD #10 the patient presented with Stage 4 DLK with UCVA of 20/100 with marked central corneal opacification, full thickness flap macrostriae and mild interface inflammation peripherally. Ocuflox® and Pred Forte® were completely discontinued and the patient was begun on Muro 128® gtt q1hr, Muro 128® ung qhs and Acular® PF QID.

On POD #15 the UCVA was decreased to 20/200 improving to 20/30- using +1.25 + 0.50 X 158. On slit lamp examination the flap folds were regressing and there was mild sub-interface haze.

At one month post op UCVA remained at 20/200 improving to 20/40- using +1.25 + 0.50 X 172. The flap macrostriae had resolved and there were focal areas of sub-interface opacification.

At 5 weeks postoperatively the flap was lifted with a Hersh spatula, irrigated with sterile BSS solution using a Dishler LASIK cannula and stretched with dry merocels®. The epithelium was left intact. At time of surgery there was no evidence for stromal necrosis, tissue loss of opacification. The flap did not demonstrate any evidence for thinning or opacity. There were no exudates, debris or any abnormal material in the interface. The patient was discharged on Vigamox® BID (Alcon, Fort Worth, TX, USA), Pred Forte® BID, Acular® PF BID, Muro 128® gtt BID and Muro 128® ung qhs.

At 10 weeks postoperatively (5 weeks post flap lift and stretch) the UCVA was 20/50 improving to 20/40 with +0.50 +0.50 X 152. Under slit lamp examination the flap and underlying stroma appeared to be normal. The patient was continued on Muro 128® gtts TID and Muro 128® ung qhs.

At 16 weeks postoperatively the UCVA was 20/25 and BCVA was also 20/25 using Plano + 0.25 X 165. Slit lamp examination appeared normal. Muro 128® was discontinued.

At 7 months postoperatively the patient presented complaining of poor vision and demonstrating a marked hyperopic shift. UCVA was reduced to 20/40 improving to 20/20 using +3.00 + 0.75 X 169. The cornea appeared normal under slit lamp examination. The patient was fitted with a +3.00 D soft contact lens (SCL). No topical medications were prescribed.

At 10 months postoperatively and after 3 months of daily SCL wear the UCVA was improved to 20/30+ and the hyperopic shift was reduced to +1.75 +1.00 X 165 providing BCVA of 20/20. The patient underwent an uncomplicated excimer laser enhancement using the VISX™ Star S4 employing a standard sphero-cylindrical ablation profile for +1.8 + 1.00 X 178. He was discharged on Ocuflox® BID, Pred Forte® QID both for one week and artificial tears q1hr. On POD #1 from the enhancement UCVA was 20/20 with a clear cornea.

At 5 months from the laser enhancement UCVA was 20/40- improving to 20/20 using a +1.75 + 0.75 X 175. A second uncomplicated excimer laser enhancement was performed using the VISX™ Star S4 employing a standard sphero-cylindrical ablation profile for +1.00 + 0.75 X 175. The patient was discharged on Zymar® (Allergan Inc, Irvine, CA, USA) and Pred Forte® QID and artificial tears q2hr.

At one-month post enhancement #2 the UCVA was 20/20 with a manifest refraction of Plano. The cornea appeared normal at slit lamp examination. The patient was discharged from the clinic.

Case 3

PJ is 43 year-old Caucasian male whose preoperative manifest refraction was OD -8.25 + 0.25 X 157 and OS -7.75 +0.50 X 65 correcting to 20/15 OD and 20/20 OS. Ultrasound pachymetry was OD 530 OD and OS 496 microns. CustomVue™ WaveScan testing and Orbscan® topography were normal OU.

Bilateral simultaneous LASIK was performed on December 17, 2003. Both LASIK flaps were created in an uncomplicated manner using an Intralase™ Femtosecond laser settings with an intended flap thickness of 90 microns and a nasal hinge, a raster energy of 2.5 microjoules on a 10 micron spot and line separation with a side cut energy of 5.0 microjoules. An uncomplicated excimer treatment was performed using a VISX™ Star S4 using a standard 6.5 mm optical zone and an 8 mm blend zone with an intended ablation depth of 110 microns OD and 103 microns OS. Postoperatively the patient was treated with Zymar®, Pred Forte® and Acular® PF QID.

On POD #1 the UCVA OD was 20/20- and 20/40 OS with 2+ interface inflammation sparing the central 2 mm visual axis of both eyes. The patient was continued on Zymar® and Pred Forte® TID.

On POD #5 the patient presented with advanced bilateral Stage 4 DLK with UCVA of 20/400 OD and 20/200 OS with bilateral marked central corneal opacification, full thickness flap macrostriae bilaterally, epithelial microcysts and bullae OD and mild interface inflammation peripherally in both eyes. Zymar® and Pred Forte® were completely discontinued and the patient was begun on Muro 128® gtt q2hr, Muro 128® ung qhs and Acular® PF QID.

On POD #7 the UCVA was 20/100 OU and the epithelial bullae OD were now resolved. No change was made to the hyperosmotic treatment regimen.

On POD #12 the UCVA was improved to 20/50+ OD but worsened to 20/200 OS. On slit lamp examination the epithelial microcysts and bullae were gone and moderate flap macrostriae persisted OD. The OS demonstrated an epithelial defect centrally, persistent central macrostriae and central corneal opacification. No change was made to the hyperosmotic treatment regimen.

On POD #14 the UCVA was improved to 20/30- OD and was unchanged at 20/200 OS. No change was made to the hyperosmotic treatment regimen.

On POD #24 the UCVA was 20/40+ OD improving to 20/20- using +0.25 + 0.50 X 3. UCVA OS was improved to 20/100 improving to 20/60- with +0.25 + 0.50 X 160. Muro 128® gtt frequency was reduced to QID and Acular was continued at QID.

At one month both flaps had the central epithelium removed using a #64 Beaver blade® and then were lifted using a Hersh spatula and irrigated with 1 cc of sterile BSS using a Dishler LASIK cannula. Both flaps were then stretched using dry merocels and bandage SCL's were inserted. The patient was discharged on Vigamox® and Pred Forte® QID for one week, Acular® PF BID, Muro 128® 128 gtts q1hr and Muro 128® ung qhs.

At day #9 post flap lift OU UCVA OD was 20/40 improving to 20/30- with -1.25 + 1.25 X 10. The OS demonstrated a marked hyperopic shift. UCVA OS was 20/60+ improving to 20/40- with +2.00 + 1.75 X 105. The patient was continued on Acular® PF BID, Muro 128® gtts q1hr and Muro 128® ung qhs.

At day #19 post flap lift OU the UCVA OD was 20/30+ improving to 20/20 with -0.25 + 1.25 X 15. The OS demonstrated a marked reduction in the hyperopic shift over the 10-day period. UCVA OS was 20/40 improving to 20/30 with -1.25. There were trace microstriae present centrally in both eyes and mild to moderate sub-interface haze OU. No change was made to the previous hyperosmotic treatment regimen.

At 7 weeks post flap lift OU the UCVA OD was 20/20- improving to 20/15- with a +1.00. UCVA OS was 20/60- improving to 20/30 using - 0.75 + 0.25 X 7. The sub-interface haze was resolving in both eyes. No change was made to the previous hyperosmotic treatment regimen.

At one year from the original LASIK procedure (11 months from flap lift and stretch) the UCVA OD was 20/20+ improving to 20/15 using -0.25 + 0.50 X 130. UCVA OS was 20/ 30 improving to 20/20 using +0.25 + 0.75 X 92. There was trace sub-interface haze in both eyes with no evidence for flap microstriae. The patient was functioning well without glasses or contact lenses and had no adverse visual symptoms. The patient elected to avoid further LASIK enhancement for the OS and was pleased with the visual outcome obtained.

Case #4

PTJ is 36 year-old Caucasian male that had previously undergone an uncomplicated LASIK procedure in the OD at our center for a manifest refraction of OD -2.50 + 1.25 X 45 using a Moria LSK-1 microkeratome (Moria-Microtek, Doylestown, PA, USA) and LADARVision® excimer laser. The OS had undergone uncomplicated cataract with intraocular lens implantation on August 3, 2003 following the LASIK in the OD. He presented with manifest refraction in the OS of -0.50 +1.25 X 97 correcting to 20/20 OD. Ultrasound pachymetry was 535 microns OS. CustomVue™ WaveScan testing and Orbscan® II topography were normal OU. Ocular history was pertinent for amblyopia OS and seasonal allergies treated using nasal steroid spray.

LASIK was performed on the OS on February 4, 2004. The LASIK flap was created in an uncomplicated manner using an Intralase™ femtosecond laser settings with an intended flap thickness of 100 microns and a nasal hinge, a raster energy of 2.5 microjoules on a 10 micron spot and line separation with a side cut energy of 4.5 microjoules. An uncomplicated excimer treatment was performed using a VISX™ Star S4 using a standard 6.5 mm optical zone and an 8 mm blend zone with an intended ablation depth of 9 microns. Postoperatively the patient was treated with Zymar®, Pred Forte® and Acular® PF QID.

On POD #1 the UCVA OS was 20/30 and there was 1+ interface inflammation superiorly and nasally. The patient was placed on Zymar® and Pred Forte® q3hrs.

On POD #8 the patient presented with advanced Stage 4 DLK with UCVA of 20/100 OS with marked central corneal opacification, full thickness flap macrostriae and mild interface inflammatory cells. The flap was immediately lifted with a Hersh spatula and irrigated with 1 cc of BSS using a Dishler LASIK cannula. At time of surgery it was observed under the surgical microscope that the stromal bed was normal with no evidence for necrosis, tissue loss or opacification. In addition, the flap did not demonstrate any evidence for thinning or opacity. There were no exudates, debris or any abnormal material in the interface. Slit lamp examination performed upon completion of the lift and irrigation procedure demonstrated absence of inflammatory cells in the interface, however the central dense opacification was unchanged from its preoperative condition. The patient was placed on Zymar®, Pred Forte® and Acular® PF QID as well as Muro 128® gtt q2hrs and artificial tears q2hrs.

On POD #10 (2 days post flap lift and irrigation) the UCVA OS was 20/70 and there was mild flap edema and sub-interface haze. Zymar® and Pred Forte® were tapered while Acular® PF was continued QID and Muro 128® was continued q2hrs.

At one month postoperatively UCVA was 20/60 improving to 20/20 with +3.00 +1.00 X 37. On slit lamp examination there were faint microstriae in the flap centrally and trace sub-interface haze. The patient was continued on Muro 128® gtts QID.

At two months postoperatively UCVA was unchanged at 20/60 improving to 20/20 with +2.25 +1.00 X 35. There was trace sub-interface haze. Muro 128® was discontinued.

At 6 months postoperatively UCVA was 20/70 improving to 20/20 with +2.00 +1.00 X 34. Orbscan® II and WaveScan measurements were within normal limits. The patient underwent an uncomplicated excimer laser enhancement using the VISX™ Star S4 employing a standard sphero-cylindrical ablation profile for +1.1 + 1.00 X 34. He was discharged on Zymar® and Pred Forte® QID both for one week and artificial tears q2hr. On POD #1 from the enhancement UCVA was 20/30 with a clear cornea.

At 4 months from the laser enhancement UCVA was 20/40- improving to 20/20 using +1.00 + 0.25 X 45. Orbscan® II and CustomVue™ WaveScan measurements were within normal limits. A second uncomplicated excimer laser enhancement was performed using the VISX™ Star S4 employing a standard sphero-cylindrical ablation profile for +0.65 + 0.25 X 23. The patient was discharged on Zymar® and Pred Forte® QID and artificial tears q2hr.

At one-month post enhancement #2 the UCVA was 20/20 with a manifest refraction of Plano + 0.25 X 41. The cornea appeared normal at slit lamp examination. The patient was discharged from the clinic.

Case #5

IM is 35 year-old Asian male whose preoperative manifest refraction was OD -3.00 + 1.75 X 72 and OS -4.25 +3.00 X 93 correcting to 20/15 OU. Preoperative keratometry OD was 40.75/42.50 @ 86 and OS 40.50/43.25 @ 96. Ultrasound pachymetry was 536 microns OD and 542 microns OS. CustomVue™ WaveScan testing and Orbscan® II topography were normal OU. The remaining ocular and medical history was unremarkable.

Bilateral simultaneous LASIK was performed on April 8, 2004. Both LASIK flaps were created in an uncomplicated manner using an Intralase™ Femtosecond laser settings with an intended flap thickness of 90 microns and a nasal hinge, a raster energy of 2.5 microjoules on a 10 micron spot and line separation with a side cut energy of 4.0 microjoules. An uncomplicated excimer treatment was performed using a VISX™ Star S4 using a standard 6.5 mm optical zone and an 8 mm blend zone with an intended ablation depth of 26 microns OD and 35 microns OS. Postoperatively the patient was treated with Zymar®, Pred Forte® and Acular® PF QID.

On POD #1 the UCVA OD was 20/20- and 20/20 OS with 1+ interface inflammation superiorly and temporally in both eyes. The Zymar® was decreased to BID and Pred Forte® continued at QID.

On POD #4 the patient presented with diffuse interface inflammation in both eyes but no evidence for flap edema. Both flaps were immediately lifted with a Hersh spatula and irrigated with 1 cc of sterile BSS using a Dishler LASIK cannula. At time of surgery it was observed under the surgical microscope that the stromal bed of both eyes was normal with no evidence for necrosis, tissue loss or opacification. In addition, neither flap demonstrated any evidence for thinning or opacity. There were no exudates, debris or any abnormal material in the interface in either eye. Slit lamp examination performed upon completion of the lift and irrigation procedure demonstrated absence of inflammatory cells in the interface.

On POD #5 (one day post bilateral flap lift and rinse) the UCVA was 20/20 OD and 20/30- OS with mild interface inflammation OS > OD in a classic "Sands of the Sahara" pattern.

On POD #6 (two days post bilateral flap lift and rinse) the UCVA was 20/20 OD and 20/60- OS. The OS now demonstrated advanced Stage 4 DLK with marked central corneal opacification, full thickness flap macrostriae and mild interface inflammation peripherally. The OS flap was immediately lifted with a Hersh spatula and irrigated with 1 cc of sterile BSS using a Dishler LASIK cannula. At time of surgery it was observed under the surgical microscope that the stromal bed was normal with no evidence for necrosis, tissue loss or opacification. In addition, the flap did not demonstrate any evidence for thinning or opacity. There were no exudates, debris or any abnormal material in the interface. Slit lamp examination performed upon completion of the lift and irrigation procedure demonstrated absence of inflammatory cells in the interface. A soft bandage contact lens was inserted in the OS. The patient was discharged on Zymar®, Pred Forte® and Acular® PF QID combined with Muro 128® gtts q1hr.

On POD #7 (three days post bilateral flap lift and rinse and 1 day post OS flap lift and rinse) the UCVA was 20/20 OD and 20/200 OS. On slit lamp examination the OD was normal post LASIK while the OS demonstrated central corneal opacification and flap macrostriae. There were no residual inflammatory cells in the interface of either eye. The Zymar® and Pred Forte® were tapered while the Muro 128® gtts were continued q1-2hrs.

On POD #20 the UCVA OS was 20/100. A drop of preservative free Glycerol (Leiter's Pharmacy, San Jose, CA, USA) was instilled in the OS X 3 over 30 minutes. Within 1 hour the UCVA was improved to 20/60-. Muro 128® gtts were continued QID.

At 5 weeks postop the UCVA OS was 20/30-2 improving to 20/20 using +0.75 + 0.75 X 78. There were faint microstriae remaining in the central flap OS as well as mild to moderate sub-interface haze.

At 6 weeks postop the central epithelium on the OS was removed using a #64 Beaver blade®, the flap was lifted with a Hersh spatula and irrigated with 1 cc of sterile BSS using a Dishler LASIK cannula. At time of surgery it was observed under the surgical microscope that the stromal bed was normal with no evidence for necrosis, tissue loss or opacification. In addition, the flap did not demonstrate any evidence for thinning or opacity. There were no exudates, debris or any abnormal material in the interface. The flap was stretched for 3 minutes using dry merocels. Slit lamp examination performed upon completion of the lift and irrigation procedure demonstrated absence of inflammatory cells in the interface. A soft bandage contact lens was inserted in the OS. The patient was discharged on Zymar®, Pred Forte® and Acular® PF QID combined with Muro 128® gtts q1hr.

On POD #2 from the epithelial removal, lift, rinse and stretch OS, the UCVA was 20/30-. There was trace sub-interface haze. The Zymar® was discontinued and the Pred Forte® continued at QID and the Muro 128® reduced to QID. Preservative free Glycerol was prescribed to be instilled 1 gtt BID OS.

On POD #8 from the flap lift and stretch OS, the UCVA was reduced to 20/60-

improving to 20/20- using +2.50 + 0.75 X 47. There was trace sub-interface haze noted. The Pred Forte® was discontinued.

On POD #19 from the flap lift and stretch the UCVA was 20/30- improving to 20/20 using +1.75 + 0.75 X 75. There was trace sub-interface haze noted. All topical medications were discontinued besides preservative free artificial tears QID.

At 6 weeks from the flap lift and stretch OS the UCVA was 20/30 improving to 20/20 using Plano + 0.75 X 79. There was persistent trace sub-interface haze noted.

At 7 months from the initial LASIK procedure the UCVA OS was 20/30- improving to 20/20 using -0.75 + 1.50 X 80. The flap and stroma were normal under slit lamp examination with no evidence for microstriae.

At 10 months from the initial LASIK procedure the UCVA was 20/30- improving to 20/20 with -0.50 +1.25 X 86. Orbscan® II and CustomVue™ WaveScan measurements were within normal limits. The patient underwent an uncomplicated excimer laser enhancement using the VISX™ Star S4 employing a standard sphero-cylindrical ablation profile for -0.75 + 1.25 X 86. He was discharged on Zymar® and Pred Forte® QID both for one week and artificial tears q2hr. On POD #1 from the enhancement UCVA was 20/20 with a clear cornea.

At one-month post excimer laser enhancement the UCVA OS was 20/20 with a manifest refraction of Plano. The cornea appeared normal at slit lamp examination. The patient was discharged from the clinic.

Case #6

LP is 43 year-old Caucasian female whose preoperative manifest refraction was OD -3.75 and OS -4.25 correcting to 20/15 OU. Ultrasound pachymetry was 498 microns OD and 481 microns OS. CustomVue™ WaveScan testing and Orbscan® II topography were normal OU. The remaining ocular and medical history was unremarkable.

Bilateral simultaneous LASIK was performed on April 8, 2004. Both LASIK flaps were created in an uncomplicated manner using an Intralase™ femtosecond laser settings with an intended flap thickness of 90 microns and a nasal hinge, a raster energy of 2.5 microjoules on a 10 micron spot and line separation with a side cut energy of 4.0 microjoules. Uncomplicated Wavefront guided excimer treatments were performed using a VISX™ Star S4 using a 6.0 mm optical zone and an 8 mm blend zone with an intended ablation depth of 56 microns OD and 63 microns OS. Postoperatively the patient was treated with Zymar®, Pred Forte® and Acular® PF QID.

On POD #1 the UCVA OD was 20/20- and 20/25- OS with 1+ interface inflammation superiorly and temporally in both eyes. The Zymar® and Pred Forte® were continued at QID.

On POD #6 the patient presented with diffuse interface inflammation in both eyes and UCVA of 20/20 OD and 20/50

OS. There was moderate central opacification of the cornea OS. Both flaps were immediately lifted with a Hersh spatula and irrigated with 1 cc of sterile BSS using a Dishler LASIK cannula. At time of surgery it was observed under the surgical microscope that the stromal bed of both eyes was normal with no evidence for necrosis, tissue loss or opacification. In addition, neither flap demonstrated any evidence for thinning or opacity. There were no exudates, debris or any abnormal material in the interface in either eye. Slit lamp examination performed upon completion of the lift and irrigation procedure demonstrated absence of inflammatory cells in the interface. Slit lamp examination performed upon completion of the lift and irrigation procedure demonstrated absence of inflammatory cells in the interface in both eyes. The sub-interface edema / opacification remained present in the OS. One drop of preservative free Glycerol was instilled in the OS. The patient was discharged on Zymar® and Acular® PF QID, Pred Forte® q2hrs and Muro 128® gtts q2hrs.

On POD #1 from the lift and rinse the UCVA was 20/20 OD and 20/50- OS. The flap and stroma were normal OD. In the OS there was moderate sub-interface haze centrally and there were mild flap microstriae present. Both interfaces were free of inflammatory cells. The Zymar® was continued at QID for one week and the Pred Forte® and Muro 128® were continued q2hrs in the OS.

On POD #14 from the flap lift and rinse OS, the UCVA was 20/40 improving to 20/30 using +1.00. There was mild sub-interface haze and mild flap microstriae noted centrally. The Pred Forte® was discontinued but Muro 128® 128 was continued q2hrs and Vitamin C was prescribed at 1000 mg PO per day.

At 8 weeks from the initial LASIK the UCVA was 20/15 OD and 20/30+ OS improving to 20/15 using +1.25 + 0.25 X 42. There was very mild sub-interface haze and mild microstriae centrally.

At 11 weeks from the initial LASIK the central epithelium on the OS was removed using a #64 Beaver blade®, the flap was lifted with a Hersh spatula and irrigated with 1 cc of sterile BSS using a Dishler LASIK cannula. At time of surgery it was observed under the surgical microscope that the stromal bed was normal with no evidence for necrosis, tissue loss or opacification. In addition, the flap did not demonstrate any evidence for thinning or opacity. There were no exudates, debris or any abnormal material in the interface. The flap was stretched for 3 minutes using dry merocels. Slit lamp examination performed upon completion of the lift and irrigation procedure demonstrated absence of inflammatory cells in the interface. A soft bandage contact lens was inserted in the OS. The patient was discharged on Zymar®, Pred Forte®, Acular® PF and Muro 128® QID.

On POD #3 from the epithelial removal, lift, rinse and stretch OS, the UCVA was 20/30-. There was no longer any detectable sub-interface haze and there were no residual flap microstriae present. The Zymar®, Pred Forte® and Acular were continued QID with a planned taper over the following two weeks.

At one month UCVA OS was 20/25 improving to 20/20 with +0.75 +0.50 X 82. At four months UCVA OS was 20/20 and at one year from the original LASIK procedure the patient demonstrated UCVA OS of 20/20 and a manifest refraction of Plano. Her corneas were normal under slit lamp examination. She was discharged from the clinic.

Bibliography

1. Smith RJ, Maloney RK,. Diffuse lamellar keratitis. A new syndrome in lamellar refractive surgery. Ophthalmology. 1998;105:1721-1726.
2. Schneider DM, Khanna R. Interface keratitis-induced stromal thinning: An early postoperative complication of laser in situ keratomileusis. J Cataract Refract Surg. 1998;24:1277-1279.
3. Lindstrom RL, Hardten DR, Houtman DM, et al. Six-month results of hyperopic and astigmatic LASIK in eyes with primary and secondary hyperopia. Trans Am Ophthalmol Soc. 1999; 97:241–260.
4. Linebarger EJ, Hardten DR, Lindstrom RL. Diffuse lamellar keratitis: Diagnosis and management. J Cataract Refract Surg. 2000;26:1072-1077.
5. Machat JJ. LASIK Complications. In Machat JJ, Slade SG, Probst LE, Eds. The Art of LASIK; 2nd Edition. Slack Inc. Thorofare, NJ. 32:392-396
6. Holland S, Morck DW, Mathias RG, Lee TL. Interface Keratitis (Sands): Clinical Update. In Probst LE, Ed. LASIK, Advances, Controversies, Custom. Slack, Inc. Thorofare, NJ. 14:175-185
7. Alio JL, Perez-Santonja JJ, Tervo T, et al. Postoperative inflammation, microbial complications, and wound healing after LASIK. Journal of Refractive Surgery. 2000;16:523-538.
8. Johnson J, Harissi-Dagher M, Pineda R, Yoo S, Azar DT. Diffuse lamellar keratitis: Incidence, association, outcomes, and a new classification system. J Cataract Refract Surg. 2001;27:1560-1566.
9. Melki SA, Azar DT. LASIK complications: Etiology, management, and prevention. Surv Ophthalmol. 2001;46(2):95-116.
10. Holland SP. Update in cornea and external disease: solving the mystery of "Sands of the Sahara" syndrome (diffuse lamellar keratitis). Can J Ophthalmol. 1999;34:193-194.
11. Mamalis N. Diffuse lamellar keratitis. J Cataract Refract Surg. 2003;29:1849-1850.
12. Stulting RD, Randleman JB, Couser JM, Thompson KP. The epidemiology of diffuse lamellar keratitis. Cornea. 2004;23:680-688.
13. Peters NT, Iskander NG, Penno EE, et al. Diffuse lamellar keratitis: Isolation of endotoxin and demonstration of the inflammatory potential in a rabbit laser in situ keratomileusis model. J Cataract Refract Surg. 2001;27:917-923.
14. Hong JW, Liu JJ, Lee J, et al. Proinflammatory chemokine induction in keratocytes and inflammatory cell infiltration into the cornea. Invest Ophthalmol Vis Sci. 2001;42:2795-2803.
15. Ambrosio RJ, Wilson S. Complications of laser in situ keratomileusis: etiology, prevention, and treatment. J Refract Surg. 2001;17:350-379.

16. Wilson SE, Ambrosio Jr R. Sporadic diffuse lamellar keratitis (DLK) after LASIK. Cornea. 2002;12(6):560-563.

17. Ambrosio Jr R, Periman LM, Netto MV, Wilson SE. Bilateral marginal sterile infiltrates and diffuse lamellar keratitis after laser in situ keratomileusis. J Refract Surg. 2003;19:154-158.

18. Asano-Kato N, Toda I, Fukomoto T, et al. Detection of neutrophils in late-onset interface inflammation associated with flap injury after laser in situ keratomileusis. Cornea. 2004;23:306-310.

19. Asano-Kato N, Toda I, Shimmura S, et al. Detection of neutrophils and possible involvement of interleukin-8 in diffuse lamellar keratitis after laser in situ keratomileusis. J Cataract Refract Surg. 2003;29:1996-2000.

20. Esquenazi S, He J, Bazan HEP, Bazan NG. Prevention of experimental diffuse lamellar keratitis using a novel platelet-activating factor receptor agonist. J Cataract Refract Surg. 2004;30:884-891.

21. Wilson SE, Mohan RR, Mohan RR, et al. The corneal wound healing response: Cytokine-mediated interaction of the epithelium, stroma, and inflammatory cells. Progress in Retinal and Eye Research. 2001;20(5):625-637.

22. Sonmez B, Maloney RK. Central toxic keratopathy: Description of a syndrome in laser refractive surgery. Am J Ophthalmol. 2007;143:420-427.

23. Hoffman RS, Fine IH, Packer M. Incidence and outcomes of LASIK with diffuse lamellar keratitis treated with topical and oral corticosteroids. J Cataract Refract Surg. 2003;29:451-456.

24. Hainline BC, Price MO, Choi DM, Price FW. Central flap necrosis after LASIK with microkeratome and femtosecond laser created flaps. J Refract Surg. 2007;23:233-242.

25. Parolini B, Marcon G, Panozzo GA. Central necrotic lamellar inflammation after laser in situ keratomileusis. J Refract Surg. 2001;17:110-112.

26. Lyle WA, Jin GJ. Central lamellar keratitis. J Cataract Refract Surg. 2001;27:487-490.

27. Holland SP, Mathias RG, Morck DW, et al. Diffuse lamellar keratitis related to endotoxins released from sterilizer reservoir biofilms. Ophthalmology. 2000;107:1227-1234.

28. Holland SP, Peters NT, Iskander NG. More to the mysterious tale: The search for the cause of 100+ cases of diffuse lamellar keratitis. J Refract Surg. 2004;20:85-86.

29. Hoffman RS, Fine IH, Packer M, et al. Surgical glove-associated diffuse lamellar keratitis. Cornea. 2005;24(6):699-704.

30. Lazaro C, Perea J, Arias A. Surgical glove-related diffuse lamellar keratitis after laser in situ keratomileusis. Long-term outcomes. J Cataract Refract Surg. 2006;32:1702-1709.

31. Fogla R. Diffuse lamellar keratitis: Are meibomian secretions responsible? J Cataract Refract Surg. 2001;27:493-495.

32. Hadden OB, McGhee CNJ, Morris AT, et al. Outbreak of diffuse lamellar keratitis caused by marking-pen toxicity. J Cataract Refract Surg 2008; 34:1121-1124.

33. Kaufman SC, Maitchouk DY, Chiou AGY, Beuerman RW. Interface inflammation after laser in situ keratomileusis. Sands of Sahara syndrome. J Cataract Refract Surg. 1998;24:1589-1593.

34. Bissen-Miyajima H, Minami K, Miyake-Kashima M, et al. Observation of the corneal flap interface with metal particles in a rabbit model. J Cataract Refract Surg. 2005;31:1409-1413.

35. Levinger S, Landau D, Kremer I, et al. Wiping microkeratome blades with sterile 100% alcohol to prevent diffuse lamellar keratitis after laser in situ keratomileusis. J Cataract Refract Surg. 2003;29:1947-1949.

36. Nakano EM, Nakano K, Oliveira MC, Portellinha W, Simonelli R, Alvarenga LS. Cleaning solutions as a cause of diffuse lamellar keratitis. J Refract Surg. 2002;18:S361-S363.

37. Yuhan KR, Nguyen L, Wachler BS. Role of instrument cleaning and maintenance in the development of diffuse lamellar keratitis. Ophthalmology. 2002;109:400-404.

38. Shen Y, Wang C, Fong S, et al. Diffuse lamellar keratitis induced by toxic chemicals after laser in situ keratomileusis. J Cataract Refract Surg. 2006;32:1146-1150.

39. Mah FS, Romanowski EG, Dhaliwal DK, et al. Role of topical fluoroquinolones on the pathogenesis of diffuse lamellar keratitis in experimental in vivo studies. J Cataract Refract Surg. 2006;32:264-268.

40. Thammano P, Rana AN, Talamo JH. Diffuse lamellar keratitis after laser in situ keratomileusis with the Moria-LSK-One and Carriazo-Barraquer microkeratomes. J Cataract Refract Surg. 2003;29:1962-1968.

41. Noda-Tsuruya T, Toda I, Asano-Kato N, et al. Risk factors for development of diffuse lamellar keratitis after laser in situ keratomileusis. J Refract Surg. 2004;20:72-75.

42. Bazan HEP. Cellular and molecular events in corneal wound healing: significance of lipid signaling. Experimental Eye Research. 2005;80:453-463.

43. Binder PS. One thousand consecutive Intralase laser in situ keratomileusis flaps. J Cataract Refract Surg. 2006;32:962-969.

44. Samuel MA, Kaufman SC, Ahee JA, Wee C, Bogorad D. Diffuse lamellar keratitis associated with carboxymethylcellulose sodium 1% after laser in situ keratomileusis. J Cataract Refract Surg. 2002;28:1409-1411.

45. Boorstein SM, Henk HJ, Elner VM. Atopy: A patient-specific risk factor for diffuse lamellar keratitis. Ophthalmology. 2003;110:131-137.

46. McLeod SD, Tham VM, Phan ST, et al. Bilateral diffuse lamellar keratitis following bilateral simultaneous versus sequential laser in situ keratomileusis. Br J Ophthalmol. 2003;87:1086-1087.

47. MacRae SM, Rich LF, Macaluso DC. Treatment of interface keratitis with oral corticosteroids. J Cataract Refract Surg. 2002;28:454-461.

48. Keszei VA. Diffuse lamellar keratitis associated with iritis 10 months after laser in situ keratomileusis. J Cataract Refract Surg. 2001;27:1126-1127.

49. Symes RJ, Catt CJ, Males JJ. Diffuse lamellar keratitis associated with gonococcal keratoconjunctivitis 3 years after laser in situ keratomileusis. J Cataract Refract Surg. 2007;33:323-325.

50. Michieletto P, Balestrazzi A. DLK and other forms of keratitis. In Buratto L, Brint S. Eds. Custom LASIK: Surgical Techniques and Complications. Slack, Inc, Thorofare, NJ. 8.9:242-245

51. Anonymous. A mysterious tale: The search for the cause of 100+ cases of diffuse lamellar keratitis. J Refract Surg. 2002;18:551-554.

52. De Rojas Silva V, Diez-Feijoo E, Rodriguez-Ares MT, Sanchez-Salorio M. Confocal microscopy of stage 4 diffuse lamellar keratitis with spontaneous resolution. J Refract Surg. 2004;20:391-396.

53. Michieletto P, Balestrazzi A, Balestrazzi A, et al. Stage 4 diffuse lamellar keratitis after laser in situ keratomileusis. Clinical, topographical, and pachymetry resolution 5 years later. J Cataract Refract Surg. 2006;32:353-356.

54. Ambrosio RJ, Wilson S. Complications of laser in situ keratomileusis: etiology, prevention, and treatment. J Refract Surg. 2001;17:350-379.

55. Wiig H, Rubin K, Reed RK. New and active role of the interstitium in control of interstitial fluid pressure: potential therapeutic consequences. Acta Anaesthesiologica Scand 2003; 47: 111-121

56. Ikema K, Inomata Y, Komohara Y, et al. Induction of matrix metalloproteinases (MMPs) and tissue inhibitors of MMPs correlates with outcome of acute experimental pseudomonal keratitis. Experimental Eye Research. 2006;83:1396-1404.

57. Bigham M, Enns CL, Holland SP, Buxton J, Patrick D, Marion S, Morck D, Kurucz M, Yuen V, Lafaille V, Shaw J, Mathias R, VanAndel M, Peck S. Diffuse lamellar keratitis complicating laser in situ keratomileusis. Post-marketing surveillance of an emerging disease in British Columbia, Canada, 2000-2002. J Cataract Refract Surg. 2005;31:2340-2344.

58. Shah MN, Misra M, Wihelmus KR, Koch Douglas. Diffuse lamellar keratitis associated with epithelial defects after laser in situ keratomileusis. J Cataract Refract Surg. 2000;26:1312-1318.

59. Peters NT, Lingua RW, Kim CH. Topical intrastromal steroid during laser in situ keratomileusis to retard interface keratitis. J Cataract Refract Surg. 1999;25:1437-1440.

60. Asano-Kato N, Toda I, Hori-Komai Y, et al. Epithelial ingrowth after laser in situ keratomileusis: Clinical features and possible mechanisms. Am J Ophthalmol. 2002;134:801-807.

61. Aldave AJ, Hollander DA, Abbott RL. Late-onset traumatic flap dislocation and diffuse lamellar inflammation after laser in situ keratomileusis. Cornea. 2002;21(6):604-607.

62. Schwartz GS, Park DH, Schloff S, Lane SS. Traumatic flap displacement and subsequent diffuse lamellar keratitis after laser in situ keratomileusis. J Cataract Refract Surg. 2001;27:781-783.

63. Ahee JA, Kaufman SC, Samuel MA, et al. Decreased incidence of epithelial defect during laser in situ keratomileusis using intraoperative nonpreserved carboxymethylcellulose sodium 0.5% solution. J Cataract Refract Surg. 2002;28:1651-1654.

64. Sachdev N, McGhee CN, Craig JP, et al. Epithelial defect, diffuse lamellar keratitis, and epithelial ingrowth following post-LASIK epithelial toxicity. J Cataract Refract Surg. 2002;28:1463-1466.

65. Asano-Kato N, Toda I, Tsubota K. Severe late-onset recurrent epithelial erosion with diffuse lamellar keratitis after laser in situ keratomileusis. J Cataract Refract Surg. 2003;29:2019-2021.

66. Esquenazi S, Bui V. Long-term refractive results of myopic LASIK complicated with intraoperative epithelial defects. J Refract Surg. 2006;22:54-60.

67. Mirshahi A, Buhren J, Kohnen T. Clinical course of severe central epithelial defects in laser in situ keratomileusis. J Cataract Refract Surg. 2004;30:1636-1641.

68. Perez-Santonja JJ, Galal A, Cardona C, et al. Severe corneal epithelial sloughing during laser in situ keratomileusis as a presenting sign for silent epithelial basement membrane dystrophy. J Cataract Refract Surg. 2005;31:1932-1937

69. Asano-Kato N, Toda I, Tsuruya T, et al. Diffuse lamellar keratitis and flap margin epithelial healing after laser in situ keratomileusis. J Refract Surg. 2003;19:30-33.

70. Weisenthal RW. Diffuse lamellar keratitis induced by trauma 6 months after laser in situ keratomileusis. J Refract Surg. 2000;16:749-751.

71. Haw WW, Manche EE. Late onset diffuse lamellar keratitis associated with an epithelial defect in six eyes. J Refract Surg. 2000;16:744-748.

72. Yeoh J, Moshegov CN. Delayed diffuse lamellar keratitis after laser in situ keratomileusis. Clinical and Experimental Ophthalmology. 2001;29:435-437.

73. Yavitz EQ. Diffuse lamellar keratitis caused by mechanical disruption of epithelium 60 days after LASIK. J Refract Surg. 2001;17:621.

74. Belda JI, Artola A, Alio J. Diffuse lamellar keratitis 6 months after uneventful laser in situ keratomileusis. J Refract Surg. 2003;19:70-71.

75. Jeng BH, Stewart JM, Hwang DG. Relapsing diffuse lamellar keratitis after laser in situ keratomileusis associated with recurrent erosion syndrome. Arch Ophthalmol. 2004;122:396-398.

76. Jin GJ, Lyle WA, Merkley KH. Late-onset idiopathic diffuse lamellar keratitis after laser in situ keratomileusis. J Cataract Refract Surg. 2005;31:435-437.

77. Pereira CR, Narvaez J, King JA. Late-onset traumatic dislocation with central tissue loss of laser in situ keratomileusis flap. Cornea. 2006;25:1107-1110.

78. Kocak I, Karabela Y, Karaman M, Kaya F. Late onset diffuse lamellar keratitis as a result of the toxic effect of Ecballium Elaterium Herb. J Refract Surg. 2006;22:826-827.

79. Javaloy J, Vidal MT, Abdelrahman AM, et al. Confocal microscopy comparison of Intralase femtosecond laser and Moria M2 microkeratome in LASIK. J Refract Surg. 2007;23:178-187.

80. Netto MV, Mohan RR, Ambrosio Jr. R, et al. Wound healing in the cornea A review of refractive surgery complications and new prospects for therapy. Cornea. 2005;24(5):509-521.

81. Javaloy J, Artola A, Vidal MT, et al. Severe diffuse lamellar keratitis after Femtosecond lamellar keratectomy [letter]. Br J Ophthalmol 2007; 91:699

82. Gil-Cazorla R, Teus MA, de Benito-Llopis L, Fuentes I. Incidence of diffuse lamellar keratitis after laser in situ keratomileusis associated with the Intralase 15 kHz Femtosecond laser and the Moria M2 microkeratome. J Cataract Refract Surg 2008; 34:28-31.

83. Vesaluoma MH, Petroll WM, Perez-Satonja JJ, et al. Laser in situ keratomileusis flap margin: Wound healing and complications imaged by in vivo confocal microscopy. Am J Ophthalmol. 2000;130:564-573.

84. Buhren J, Baumeister M, Kohnen T. Diffuse lamellar keratitis after laser in situ keratomileusis imaged by confocal microscopy. Ophthalmology. 2001;108:1075-1081.

85. Harrison DA, Periman LM. Diffuse lamellar keratitis associated with recurrent corneal erosions after laser in situ keratomileusis. J Refract Surg. 2001;17:463-465.

86. Chung MS, Pepose JS, Al-Agha S, Cavanagh HD. Confocal microscopic findings in a case of delayed-onset diffuse lamellar keratitis after laser in situ keratomileusis. J Cataract Refract Surg. 2002;28:1467-1470.

87. Buhren J, Baumeister M, Cichocki M, Kohnen T. Confocal microscopic characteristics of stage 1 to 4 diffuse lamellar keratitis after laser in situ keratomileusis. J Cataract Refract Surg. 2002;28:1390-1399.

88. DeRojas Silva V, Rodriguez-Ares T, Dies-Feijoo E, Sanchez-Salorio M. Confocal microscopy in late-onset diffuse lamellar keratitis after laser in situ keratomileusis. Ophthalmic Surgery Lasers & Imaging. 2003;34(1):68-72.

89. Holzer MP, Solomon KD, Vroman DT, et al. Diffuse lamellar keratitis: Evaluation of etiology, histopathologic findings, and clinical implications in an experimental animal model. J Cataract Refract Surg. 2003;29:542-549.

90. Moilanen JA, Holopainen JM, Helinto M, et al. Keratocyte activation and inflammation in diffuse lamellar keratitis after formation of an epithelial defect. J Cataract Refract Surg. 2004;30:341-349.

91. DeRojas Silva V, Abraldes MJ, Diez-Feijoo E, et al. Confocal microscopy and histopathological examination of diffuse lamellar keratitis in an experimental animal model. J Refract Surg. 2007;23:299-304.

92. Kurian M, Shetty R, Shetty BK, et al. In vivo microscopic findings of interlamellar stromal keratopathy induced by elevated intraocular pressure. J Cataract Refract Surg. 2006;32:1563-1566.

93. Dada T, Pangtey M, Sharma N, et al. Hyperopic shift after LASIK induced Diffuse lamellar keratitis. BMC Ophthalmology 2006; 6:19

94. Coleman DJ, Dallow RL. Introduction to Ophthalmic Ultrasonography. Ed. Duane TD. Clinical Ophthalmology. 1985;25:1-5

95. Bazan HEP, Birkle DL, Beuerman R, Bazan NG. Cryogenic lesion alters the metabolism of arachidonic acid in rabbit cornea layers. Invest Ophthalmol Vis Sci. 1985;26:474-480.

96. Bazan HEP. The synthesis and effects of Eicosanoids in avascular ocular tissues. Prog. Clin. Biol. Res. 1989;312:73–84.

97. Bazen HE, Birkle DL, Beuerman RW, Bazan NG. Inflammation-induced stimulation of the synthesis of prostaglandins and lipoxygenase-reaction products in rabbit cornea. Current Eye Research. 1985;4(3):175-179.

98. Wilson SE, Chaurasia SS, Medeiros FW. Apoptosis in the initiation, modulation and termination of the corneal wound healing response. Experimental Eye Research. 2007:1-7.

99. Mohan RR, Hutcheon AEK, Choi R, et al. Apoptosis, necrosis, proliferation and myofibroblast generation in the stroma following LASIK and PRK. Experimental Eye Research. 2003;76:71-87.

100. Dupps Jr WJ, Wilson SE. Biomechanics and wound healing in the cornea. Experimental Eye Research. 2006;83:709-720.

101. Maurice DM. The structure and transparency of the cornea. J Physiol (Lond). 1957;136:263

102. Maurice DM. The cornea and sclera. In H. Davson Ed. Vegetative Physiology and Biochemistry (3rd ed.) Academic, NY. 1984;49-57

103. Edelhauser HF, Geroski DH, Ubels JL. Physiology. In. Smolin G and Thoft RA Eds. The Cornea (3rd Ed.) 1994; 2:25-46

104. Meyer FA. Macromolecular basis of glomerular protein exclusion in loose connective tissue and of swelling pressure (umbilical cord). Biochim Biophys Acta 1983; 755: 388-399

105. Lund T, Wiig H, Reed RK. Acute post burn edema: role of strongly negative interstitial fluid pressure. Am J Physiol 1988; 255: H1069-H1074

106. Lund T, Onarheim H, Reed RK. Pathogenesis of edema formation in burn injuries. World J Surg 1992; 16: 2-9

107. Lund T, Onarheim H, Wiig H, Reed RK. Mechanisms behind increased dermal imbibition pressure in acute burn edema. Am J Physiol 1989; 256: H940-H948

108. Kinsky MP, Guha SC, Button BM, Kramer GC. The role of interstitial starling forces in the pathogenesis of burn edema. J Burn Care Rehabil 1998; 19:1-9

109. Shimizu S, Tanaka H, Sakaki S, et al. Burn depth effects dermal interstitial pressure, free radical production, and serum histamine levels in rats. J Trauma 2002; 52: 683-687

110. Koller ME, Reed RK. Increased negativity of interstitial fluid pressure in rat trachea in dextran anaphylaxis. J Appl Physiol 1992; 72: 53-57

111. Koller ME, Woie K, Reed RK. Increased negativity of interstitial fluid pressure in rat trachea after mast cell degranulation. J Appl Physiol 1993; 74: 2135-2139

112. Woie K, Koller ME, Heyeraas KJ, Reed RK. Neurogenic inflammation in rat trachea is accompanied by increased negativity of interstitial fluid pressure. Circ Res 1993; 73: 839-845

113. Rothschild AM, Gomes EL, Rossi MA. Reversible rat mesenteric mast cell swelling caused by vagal stimulation or sham-feeding. Agents Actions 1991; 34: 295-301

114. Karp G. Interactions between cells and their environment. In Karp G, Ed. Cell and Molecular Biology: Concepts and Experiments. 5th Edition John Wiley and Sons, Inc. 2008; 7:239-273

115. Hynes RO. Integrins: versatility, modulation, and signaling in cell adhesion. Cell 1992; 69: 11-25

116. Miranti CK, Brugge JS. Sensing the environment: a historical perspective on integrin signal transduction. Nat Cell Biol 2002; 4: E83-E90

117. Gullberg DE, Lundgren-Akerlund E. Collagen-binding I domain integrins – what do they do? Prog Histochem Cytochem 2002; 37: 3-54

118. Gullberg DE, Gehlsen KR, Turner DC, et al. Analysis of $\alpha_1\beta_1$, $\alpha_1\beta_1$ and $\alpha_1\beta_1$ integrins in cell-collagen interactions: identification of conformation dependent $\alpha_1\beta_1$ binding sites in collagen type I. EMBO J 1992; 11: 3865-3873

119. Wang N, Butler JP, Ingber DE. Mechanotransduction across the cell surface and through the cytoskeleton. Science 1993; 260: 1124-1127

120. Sundberg C, Rubin K. Stimulation of β_1 integrins on fibroblasts induces PDGF independent tyrosine phosphorylation of PDGF beta-receptors. J Cell Biol 1996; 132; 741-752

121. Schwartz MA, Ginsberg MH. Networks and crosstalk: integrin signaling spreads. Nat Cell Biol 2002; 4: E65-E68

122. Reed RK, Berg A, Gjerde EA, Rubin K. Control of interstitial fluid pressure: role of beta-1 integrins. Semin Nephrol 2001; 21: 222-230

123. Bell E, Ivarsson B, Merrill C. Production of a tissue-like structure by contraction of collagen lattices by human fibroblasts of different proliferative potential in vitro. Proc Natl Acad Sci USA 1979; 76: 1274-1278

124. Grinnell F. Fibroblasts, myofibroblasts, and wound contraction. J Cell Biol 1994; 124: 401-404

125. Harris AK, Stopak D, Wild P. Fibroblast traction as a mechanism for collagen morphogenesis. Nature 1981; 290: 249-251

126. Balaban NQ, Schwarz US, Riveline D, et al. Force and focal adhesion assembly: a close relationship studied using elastic micropatterned substrates. Nat Cell Biol 2001; 3: 466-472

127. Gullberg D, Tingstrom A, Thuresson AC et al. β_1 integrin-mediated collagen gel contraction is stimulated by PDGF. Exp Cell Res 1990; 186: 264-272

128. Clark RA, Folkvord JM, Hart CE, Murray MJ, McPherson JM. Platelet isoforms of platelet-derived growth factor stimulate fibroblasts to contract collagen matrices. J Clin Invest 1989; 84: 1036-1040

129. Klein CE, Dressel D, Steinmayer T, et al. Integrin $\alpha_2\beta_1$ is upregulated in fibroblasts and highly aggressive melanoma cells in three-dimensional collagen lattices and mediates the reorganization of collagen I fibrils. J Cell Biol 1991; 115: 1427-1436

130. Schiro JA, Chan BM, Roswit WT, et al. Integrin $\alpha_2\beta_1$ (VLA-2) mediates reorganization and contraction of collagen matrices by human cells. Cell 1991; 67: 403-410

131. Guidry C, Grinnell F. Studies on the mechanism of hydrated collagen gel reorganization by human skin fibroblasts. J Cell Sci 1985; 79: 67-81

132. Rubin K, Sundberg C, et al. Integrins: transmembrane links between the extracellular matrix and cell interior. In: Reed RK, McHale NG, Bert JL, Winlove CP, Laine GA (Eds) Interstitium, Connective Tissue and Lymphatics. London: Portland Press, 1995: pp 29-40

133. Reed RK, Rubin K, Wiig H, Rodt SA. Blockade of β_1 integrins in skin causes edema through lowering of interstitial fluid pressure. Circ Res 1992; 71: 978-983

134. Berg A, Rubin K, Reed RK. Cytochalasin D induces edema formation and lowering of interstitial fluid pressure in rat dermis. Am J Physiol Heart Circ Physiol 2001; 281: H7-H13

135. Bronstad A, Reith A, Berg A, Reed RK. Effect of the cytoskeletal fixation agent phalloidin on transcapillary albumin transport and interstitial fluid pressure in anaphylaxis in the Wistar rat. Microcirculation 2002; 9: 197-205

136. Erhlich HPW, Rockwell WB, Cornwell TL, Rajarathnam JB. Demonstration of a direct role for myosin light chain kinase in fibroblast-populated collagen lattice contraction. J Cell Physiol 1991; 146: 1-7

137. Gillery P, Cousty F, Pujol JP, Borel JP. Inhibition of collagen synthesis by interleukin-1 in three dimensional collagen lattice cultures of fibroblasts. Experientia 1989; 45: 98-101

138. Tingstrom A, Heldin CH, Rubin K. Regulation of fibroblast-mediated collagen gel contraction by platelet-derived growth factor, interleukin-1 alpha and transforming growth factor-beta.1. J Cell Sci 1992; 102: 315-322

139. Berg A, Hultgard-Ekwall AK, Rubin K, Stjernschantz J, Reed RK. Effect of PGE_1, PGI_2, and $PGF_{2\alpha}$ analogs on collagen gel compaction in vitro and interstitial pressure in vivo. Am J Physiol 1998; 274: H663-H671

140. Salnikov AV, Iversen VV, Koisti M, Sundberg C, Johansson L, Stuhr LB, Sjoquist M, Ahlstrom H, Reed RK, Rubin K. Lowering of tumor interstitial fluid pressure specifically augments efficiency of chemotherapy. FASEB J 2003; 17: 1756-1758

141. Rubin K, Sjoquist M, Gustafsson AM, Isaksson B, Salvessen G, Reed RK. Lowering of tumoral interstitial fluid pressure by prostaglandin E(1) is paralleled by an increased uptake of (51)Cr-EDTA. Int J Cancer 2000; 86: 636-643

142. Iversen VV, Reed RK. PGE1 induced transcapillary transport of 51Cr-EDTA in rat skin measured by microdialysis. Acta Physiol Scand 2002; 176: 269-274

143. Iversen VV, Bronstad A, Gjerde EA, Reed RK. Continuous measurements of plasma protein extravasation with microdialysis following various inflammatory challenges in rat and mouse skin. Am J Physiol Heart Circ Physiol 2004; 286:H108-112

144. Nedrebo T, Berg A, Reed RK. Effect of tumor necrosis factor-alpha, IL-1beta and IL-6 on interstitial fluid pressure in rat skin. Am J Physiol 1999; 277: H1857-1862

145. Iversen VV, Borge NBA, Salvesen GS, Reed RK. Platelet activation factor (PAF) increases plasma protein extravasation and induces lowering of interstitial fluid pressure (P_{if}) in rat skin. Acta Physiol Scand. 2005;185:5-12.

146. Ahlen K, Berg A, Stiger F, et al. Cell interactions with collagen matrices in vivo and in vitro depend on phophatidylinositol 3-kinase and free cytoplasmic calcium. Cell Adhes Commun 1998; 5: 461-473

147. Lund T, Reed RK. α-Trinositol inhibits edema generation and albumin extravasation in thermally injured skin. J Trauma 1994; 36: 761-765

148. Reed RK, Westerberg EJ. Effect of α-Trinositol on carrageenan-induced rat paw edema and lowering of interstitial fluid pressure. Eur J Pharmacol 1999; 376: 279-284

149. Rodt SA, Reed RK, Ljungstrom M, Gustafsson TO, Rubin K. The anti-inflammatory agent α-Trinositol exerts its edema-preventing effects through modulation of β-1 integrin function. Circ Res 1994; 75: 942-948

150. Berg A, Aas P, Gustafsson T, Reed RK. Effect of α-Trinositol on interstitial fluid pressure, oedema generation and albumin extravasation in experimental frostbite in the rat. Br J Pharmacol 1999; 126: 1367-1374

151. Woie K, Westerberg E, Reed RK. Lowering of interstitial fluid pressure will enhance edema in the trachea of albumin-sensitized rats. Am J Respir Crit Care Med 1996; 153: 1347-1352

152. Woie K, Reed RK. Neurogenic inflammation and lowering of interstitial fluid pressure in rat trachea is inhibited by α-Trinositol. Am J Respir Crit Care Med 1994; 150: 924-928

153. Rodt SA, Ahlen K, Berg A, Rubin K, Reed RK. A novel physiological function for platelet-derived growth factor-BB in rat dermis. J Physiol (London) 1996; 495: 193-200

154. Liden A, Berg A, Nedrebo T, et al. Platelet-derived growth factor BB-mediated normalization of dermal interstitial fluid pressure after mast cell degranulation depends on $\beta 3$ but not $\beta 1$ integrins. Circ Res. 2006;98:635-641.

155. Heuchel R, Berg A, Tallquist M, et al. Platelet-derived growth factor β receptor regulates interstitial fluid homeostasis through phosphatidylinositol-3' kinase signaling. Proc Natl Acad Sci USA 1999; 96: 11410-11415

156. Bron AJ. The architecture of the corneal stroma. Br J Ophthalmol. 2001;85:379-381.

157. Muller LJ, Pels E, Vrensen GF. The specific architecture of the anterior stroma accounts for maintenance of corneal curvature. Br J Ophthalmol. 2001;85:437-443.

158. Edelhauser HF. The balance between corneal transparency and edema. The proctor lecture. Invest Ophthalmol Vis Sci 2006;47(5):1755-1767.

159. Grinnell F. Fibroblast biology in three-dimensional collagen matrices. Trends in Cell Biology. 2003;13(5):264-269.

160. Petroll WM, Vishwanath M, Ma L. Corneal fibroblasts respond rapidly to changes in local mechanical stress. IVOS. 2004;45(1):3466-3474.

161. Vishwanath M, Ma L, Otey CA, et al. Modulation of corneal fibroblast contractility within fibrillar collagen matrices. Invest Ophthalmol Vis Sci. 2003;44(11):4724-4730.

162. Roy P, Petroll WM, Chunong CJ, et al. Effect of cell migration on the maintenance of tension on a collagen matrix. Annals of Biomedical Engineering. 1999;27:721-730.

163. Ruberti JW, Klyce SD, Smolek MK, Karon MD. Anomalous acute inflammatory response in rabbit corneal stroma. Invest Ophthalmol Vis Sci. 2000;41(9):2523-2530.

164. Karon MD, Klyce SD. Effect of inhibition of inflammatory mediators on trauma-induced stromal edema. Invest Ophthalmol Vis Sci. 2003;44(6): 2507-2511.

165. Wigham CG, Turner HC, Swan J, Hodson SA. Modulation of corneal endothelial hydration control mechanisms by Rolipram. Eur J Physiol. 2000;440:866-870.

166. Wiig H. Cornea fluid dynamics I: Measurement of hydrostatic and colloid osmotic pressure in rabbits. Exp Eye Res. 1989;49:1015-1030.

167. Wiig H. Cornea fluid dynamics II. Evidence for transport of radiolabelled albumin in rabbits by bulk flow. Exp Eye Res. 1990;50(3):261-7.

168. Amann J, Holley G, Lee SB, Edelhauser HF. Increased endothelial cell density in the paracentral and peripheral regions of the human cornea. Am J Ophthalmol. 2003;135:584-590.

169. Moller-Petersen T, Ehlers N. A three-dimensional study of the human corneal keratocyte density. Curr Eye Res. 1995;14(6):459-64.

170. Patel S, McLaren J, Hodge D, Bourne W. Normal human keratocyte density and corneal thickness measurement by using confocal microscopy in vivo. Invest Ophthalmol Vis Sci. 2001;42(2):333-9.

171. Hahnel C, Somodi S, Weiss DG, Guthoff RF. The keratocyte network of human cornea: a three dimensional study using confocal laser scanning fluorescence microscopy. Cornea. 2000;19(2):185-93.

172. Netto MV, Wilson SE. Corneal wound healing relevance to wavefront guided laser treatments. Ophthalmol Clin N Am. 2004;17:225-231.

173. Kurtz RM, Sarayba MA, Juhasz T. Intralase: Clinical Update. In Probst LE, Ed. LASIK, Advances, Controversies, Custom. Slack, Inc. Thorofare, NJ. 19:231-7.

174. Hu MY, McCulley JP, Cavanagh HD, Bowman RW, Verity SM, Mootha VV, Petroll WM. Comparison of the corneal response to laser in situ keratomileusis with flap creation using the FS15 and FS30 femtosecond lasers; clinical and confocal microscopy findings. J Cataract Refract Surg. 2007;33(4):673-81.

175. Kim JY, Kim MJ, Kim TI et al. A femtosecond laser creates a stronger flap than a mechanical microkeratome. Invest Ophthalmol Vis Sci. 2006;47:599-604.

176. Netto MV, Mohan RR, Medeiros FW et al. Femtosecond laser and microkeratome corneal flaps: comparison of stromal wound healing and inflammation. J Refract Surg 2007;23(7):667-76.

177. Alon R, Dustin ML. Force as a facilitator of integrin conformational changes during leukocyte arrest on blood vessels and antigen-presenting cells. Immunity. 2007;26(1):17-27

178. Zhang X, Chen A, De Leon D et al. Atomic force microscopy measurement of leukocyte-endothelial interaction. Am J Physiol Heart Circ Physiol. 2004;286(1):H359-67.

179. Wang SK, Chiu JJ, Lee MR et al. Leukocyte-endothelium interaction: measurement by laser tweezers force spectroscopy. Cardiovasc Eng. 2006;6(3):111-7.

180. Chang-Godinich A, Steinert RF, Wu HK. Late occurrence of diffuse lamellar keratitis after laser in-situ keratomileusis. Arch Ophthalmol. 2001;119:1074-6.

181. Steinert RF, McColgin AZ, White A, Horsburgh GM. Diffuse interface keratitis after laser in situ keratomileusis (LASIK): a nonspecific syndrome. Am J Ophthalmol. 2000;129(3):380-1.

182. Gris O, Guell JL, Wolley-Dod C, et al. Diffuse lamellar keratitis and corneal edema associated with viral keratoconjunctivitis 2 years after laser in situ keratomileusis. J Cataract Refract Surg. 2004;30:1366-1370.

183. Perez-Santonja JJ, Linna TM, Tervo KM et al. Corneal wound healing after laser insitu keratomileusis in rabbits. J Refract Surg. 1998;14:602-9.

184. Roberts C. The cornea is not a piece of plastic. J Refract Surg. 2000;16:407-13.

185. Roberts C. The cornea is not a piece of plastic. Letters to the editor. J Refract Surg. 2000;17:76-8.

186. Roberts C. Characterization of corneal curvature changes inside and outside the ablation zone in LASIK. Invest Ophthalmol Vis Sci. 2000;41(suppl):S679.

187. Dupps Jr WJ, Wilson SE. Biomechanics and wound healing in the cornea. Experimental Eye Research. 2006;83:709-720.

188. Wright DM, Wiig H, Winlove P et al. Simultaneous measurement of interstitial fluid pressure and load in rat skin after strain application in vitro. Annals of Biomedical Eng. 2003;31:1246-54.

189. Kim KS, Jean SJ, Edelhauser HF. Corneal endothelial morphology and barrier function following excimer laser photorefractive keratectomy. In Lass J, et al., (eds) Advances in Corneal Research. New York; Plenum Press. 1997 pp 329-42

190. Kim T, Sorenson AL, Krishnasamy S, Carlson AN, Edelhauser HF. Acute corneal endothelial changes after laser in situ keratomileusis. Cornea 2001; 20:597-602

191. Edelhauser HF. The resiliency of the corneal endothelium to refractive and intraocular surgery. Cornea 2000; 19:263-73

192. Carones F, Brancato R, Venturi E, Vigo L. The human corneal endothelium after myopic excimer laser photorefractive keratectomy; Immediate to one-month follow-up. Eur J Ophthalmol 1995; 5:204-13

193. Carones F, Brancato R, Venturi, Morico A. The corneal endothelium after myopic excimer laser photorefractive keratectomy. Arch Ophthalmol 1994; 112:920-4

194. Mardetti RG, Piebenga LW, Matta CS, et al. Corneal endothelial status 12 to 55 months after excimer laser photorefractive keratectomy. Ophthalmology 1995; 102:544-9

195. Amano S, Shimizu K. Corneal endothelial changes after excimer laser photorefractive keratectomy. Am J Ophthalmol 1993; 116:692-4

196. Rosa N, Cennamo G, Del Prete A, et al. Effects on the corneal endothelium 6 months following photorefractive keratectomy. Ophthalmologica 1995; 2098:17-20

197. Perez-Santonja JJ, Meza J, Moreno E, et al. Short-term corneal endothelial changes after photorefractive keratectomy. J Refract Corneal Surg 1994; 10:S194-8

198. Jones SS, Azar RG, Cristol SM, et al. Effects of laser in situ keratomileusis (LASIK) on the corneal endothelium. Am J Ophthalmol 1998; 125:465-71

199. Perez-Santonja JJ, Sakla HF, Alio JL. Evaluation of endothelial cell changes one year after excimer laser in situ keratomileusis. Arch Ophthalmol 1997;115:841-6

200. Perez-Santonja JJ, Sakla HF, Gobbi F, et al. Corneal endothelial changes after laser in situ keratomileusis. J Cat Ref Surg 1997; 23:177-83

201. Krueger RR, Thornton IL, Xu M et al. Rainbow glare as an optical side effect of IntraLASIK. Ophthalmology. 2008;115(7):1187-95.
202. Stoy V, Fernandes M. The cornea is not a piece of plastic. Letters to the editor. J Refract Surg. 2000;17:76-8.
203. Stoy VA. Hydrogels. In, Swarbrick J, Boylan JC, Eds. Encyclopedia of Pharmaceutical Technology. Vol 18 supplement. New York, NY. Marcel Dekker, Inc. 1999:91-119.
204. Nordlund ML, Grimm S, Lane S, Holland E. Pressure-induced interface keratitis: A late complication following Lasik. Cornea. 2004;23:225-234.
205. Lyle WA, Jin GJ. Interface fluid associated with diffuse lamellar keratitis and epithelial ingrowth after laser in situ keratomileusis. J Cataract Refract Surg. 1999;25:1009-1012.
206. Galal A, Artola A, Belda J, et al. Interface corneal edema secondary to steroid-induced elevation of intraocular pressure simulating diffuse lamellar keratitis. J Refract Surg. 2006;22:441-447.
207. Cheng AC, Law RW, Young AL, Lam DS. In vivo confocal microscopic findings in patients after steroid-induced glaucoma after LASIK. Ophthalmology. 2004;111:768-774.
208. Dawson DG, Schmack I, Holley GP et al. Interface fluid syndrome in human eye bank corneas after LASIK: Causes and Pathogenesis. Ophthalmology 2007;114:1848-59.
209. Fam H, Goh E, Lee H et al. Post LASIK myopic shift after a trek in the North Pole. J Cataract Refract Surg 2005;31:198-201
210. White LJ, Mader TH. Refractive changes at high altitude after LASIK [letter] Ophthalmology 2000;107:2118
211. Boes DA, Omura AK, Hennessy MJ. Effects of high-altitude exposure on myopic laser in situ keratomileusis. J Cataract Refract Surg 2001;27:1937-41.
212. Karp G. Cell Signaling and Signal Transduction: Communication between Cells. In Karp G, Ed. Cell and Molecular Biology: Concepts and Experiments. 5th Edition John Wiley and Sons, Inc. 2008; 17:616-61.
213. Erie JC, Patel SV, McLaren JW et al. Corneal keratocyte deficits after photorefractive keratectomy and laser in situ keratomileusis. Am J Ophthalmol 2006;141(5):799-809.
214. Erie JC, McLaren JW, Hodge DO and Bourne WM. Long-term corneal keratocyte deficits after photorefractive keratectomy and laser in situ keratomileusis. Trans Am Ophthalmol Soc 2005;103:56-66.
215. Machemer R. Vitrectomy – a pars plana approach. In Machemer and Benson, (Eds) Grune and Stratton, New York 1975; pp 51
216. Edelhauser HF, Hine JE, Pederson H, Van Horn DL, Schultz RO. The effect of phenylephrine on the cornea. Arch Ophthalmol 1979; 97:937-47
217. Antoine ME, Edelhauser HF, O'Brien WJ. Pharmokinetics of topical ocular phenylephrine HCl. Invest Ophthalmol Vis Sci 1984; 25:48-54

218. Staatz WD, Van Horn DL, Edelhauser HF, Schultz RO. Effects of phenylephrine on bovine corneal endothelium in culture. Ophthalmic Res 1980; 12:244

219. Lapalus P, Ettaiche M, Fredj-Reygrobellet D, Jambou D, Elena PP. Cytotoxicity studies in ophthalmology. Lens Eye Tox Res 1990; 7:231-242

220. Walkenbach RJ, Ye GS, Reinach PS, Boney F. Alpha$_1$-adrenoceptors in the corneal endothelium. Exp Eye Res. 1992; 55(3):443-50

221. Berridge MJ. Inositol triphosphate and diacylglycerol: two interacting second messengers. Ann Rev Biochem 1987; 56:159-93

222. Minneman KP. Alpha$_1$-adrenergic receptor subtypes, inositol phosphates, and sources of cell Ca^{2+}. Pharmacol Rev 1988; 40:87-112

223. Partington CR, Edwards MW, Daly JW. Regulation of cyclic AMP formation in brain tissue by α_1-adrenergic receptors: requisite intermediacy of prostaglandins of the E series. Proc Nat Acad Sci USA 1980; 77:3024-28

224. Schaad NC, Schorderet M, Magistretti PJ. Prostaglandins and the synergism between VIP and noradrenaline in the cerebral cortex. Nature 1987; 328:637-40

225. Walkenbach RJ, Ye GS, Reinach PS, Boney F. Alpha 1-adrenoceptors in human corneal epithelium. Inv Ophthalmol Vis Sci 1991; 32: 3067-3072

226. Engstrom P, Dunham EW. Alpha-adrenergic stimulation of prostaglandin release from rabbit iris-ciliary body in vitro. Invest Ophthal Vis Sci 1982; 22: 757-767

227. Yohai D, Danon A. Effect of adrenergic agonists on eiconsanoid output from isolated rabbit choroid plexus and iris-ciliary body. Prostaglandins Leukot Med 1987; 28: 227-235

228. Yousufzai SY, Abdel-Latif AA. Alpha-1-adrenergic receptor induced subsensitivity and supersensitivity in rabbit iris-ciliary body. Effects on myo-inositol triphosphate accumulation, arachidonate release, and prostaglandin synthesis. Invest Ophthalmol Vis Sci 1987; 28:409-419

229. Abdel-Latif AA, Smith JP. Studies on the incorporation of [1-14C] arachidonic acid into glycerolipids and its conversion into prostaglandins by rabbit iris. Effects of anti-inflammatory drugs and phospholipase A2 inhibitors. Biochim Biophys Acta 1982; 11: 478-489

230. Miyake K, Shirasawa E, Hikita M, Miyake Y, Kuratomi R. Synthesis of prostaglandin E in rabbit eyes with topically applied epinephrine. Invest Ophthalmol Vis Sci 1988; 29: 332-334

231. Walter KA, Gilbert DD. The adverse effect of perioperative Brimonidine tartrate 0.2% on flap adherence and enhancement rates in Laser In Situ Keratomileusis patients. Ophthalmology 2001; 108:1434-8

232. Machat JJ. LASIK Complications. In Machat JJ, Slade SG, Probst LE, Eds. The Art of LASIK; 2nd Edition. Slack Inc. Thorofare, NJ. 32:409-411.

233. Probst LE. Myopic and Hyperopic LASIK Complications. In Probst LE, Ed. LASIK, Advances, Controversies, Custom. Slack, Inc. Thorofare, NJ. 12:162-3.

234. Peng Q, Holzer MP, Kaufen PH, Apple DJ, Solomon KD. Interface fungal after laser in situ keratomileusis presenting as diffuse lamellar keratitis. A clinicopathological report. J Cataract Refract Surg. 2002;28:1400-1408.

235. Tanaka H, Lund T, Wiig H, Reed RK, Yukioka T, Matsuda H, Shimazaki S. High dose vitamin C counteracts the negative interstitial fluid hydrostatic pressure and early edema generation in thermally injured rats. Burns 1999; 25: 569-574

236. Williams RN, Paterson CA. Modulation of corneal lipoxygenase by ascorbic acid. Exp Eye Res 1986; 43: 7-13

237. Kasetsuwan N, Wu FM, Hsieh F, Sanchez D, McDonnell PJ. Effect of topical ascorbic acid on free radical tissue damage and inflammatory cell influx in the cornea after excimer laser corneal surgery. Arch Ophthal 1999; 117: 649-652

238. Stojanovic A, Ringvold A, Nitter T. Ascorbate prophylaxis for corneal haze after photorefractive keratectomy. J Ref Surg 2003; 19: 338-343

239. Cosar CB, Sener AB, Sen N, Coskunseven E. The efficacy of hourly prophylactic steroids in diffuse lamellar keratitis epidemic. Ophthalmologica. 2004;218(5):318-322.

240. Holzer MP, Soloman KD, Vargas LG et al. Diffuse lamellar keratitis. Postoperative prophylactic treatment with corticosteroids in an experimental animal study. Ophthalmologe. 2002;99(11):849-53.

241. Sandoval HP, Vargas LG, Holzer MP et al. Diffuse lamellar keratitis: prophylactic treatment with ketorolac tromethamine 0.5% in an animal model. Arch Soc Esp Oftalmol. 2002;77(11):589-95.

242. Holzer MP, Sandoval HP, Vargas LG, et al. Corneal flap complications in refractive surgery. Part 2: Postoperative treatments of diffuse lamellar keratitis in an experimental animal model. J Cataract Refract Surg. 2003;29:803-807.

243. Holzer MP, Sandoval HP, Vargas LG, et al. Evaluation of preoperative and postoperative prophylactic regimens for prevention and treatment of diffuse lamellar keratitis. J Cataract Refract Surg. 2004;30:195-199.

244. Morck DW, Holland SP, Ceri H, et al. Use of polymyxin as an endotoxin blocker in the prevention of diffuse lamellar keratitis in an animal model. J Refract Surg. 2005;21:152-157.

245. Stuhr LE, Reith A, Lepsoe S, Myklebust R, Wiig H, Reed RK. Fluid pressure in human dermal fibroblast aggregates measured with micropipettes. Am J Physiol Cell Physiol 2003;285(5):C1101-8.

246. Davidson RS, Brandt JD, Mannis MJ. Intraocular pressure-induced interlamellar keratitis after LASIK surgery. J Glaucoma. 2003;12(1):23-26.

247. Leu G, Hersh PS. Phototherapeutic keratectomy for the treatment of diffuse lamellar keratitis. J Cataract Refract Surg. 2002;28:1471-1474.

248. Yoon G, MacRae S, Williams D et al. Causes of spherical aberration induced by laser refractive surgery. J Cataract Refract Surg. 2005;31:127-135.

249. Vinciguerra P, Camesasca FI, Torres IM. Transition zone design and smoothing in custom laser-assisted subepithelial keratectomy. J Cataract Refract Surg 2005;31:39-47.

250. Qazi MA, Roberts CJ, Mahmoud AM et al. Topographic and biomechanical differences between hyperopic and myopic laser in situ keratomileusis. J Cataract Refract Surg 2005;31:48-60.